Linux for Embedded and Real-Time Applications

Linux for Embedded and Real-Time Applications

SECOND EDITION

by Doug Abbott

AMSTERDAM • BOSTON • HEIDELBERG • LONDON
NEW YORK • OXFORD • PARIS • SAN DIEGO
SAN FRANCISCO • SINGAPORE • SYDNEY • TOKYO

Newnes is an imprint of Elsevier

Newnes is an imprint of Elsevier
30 Corporate Drive, Suite 400, Burlington, MA 01803, USA
Linacre House, Jordan Hill, Oxford OX2 8DP, UK

 Recognizing the importance of preserving what has been written, Elsevier prints its books on acid-free paper whenever possible.

Library of Congress Cataloging-in-Publication Data

Abbott, Doug.
Linux for embedded and real-time appplications / Doug Abbott. – 2nd ed.
p. cm.
ISBN 0-7506-7932-8 (pbk. : alk. paper) 1. Linux. 2. Operating systems (Computers)
3. Embedded computer systems—Programming. 4. Real-time programming. I. Title.
QA76.76.O63A24 2006
005.4'32--dc22

2006005262

British Library Cataloguing-in-Publication Data
A catalogue record for this book is available from the British Library.

ISBN-13: 978-0-7506-7932-9
ISBN-10: 0-7506-7932-8

For information on all Newnes publications, visit our Website at www.books.elsevier.com

06 07 08 09 10 9 8 7 6 5 4 3 2 1

Printed in the United States of America.

Dedication

To Susan: My best friend, my soul mate.
Thanks for sharing life's journey with me.

To Brian: Budding actor/musician/lighting director,
future pilot and all around neat kid.
Thanks for keeping me young at heart.

Contents

Preface

" 'You are in a maze of twisty little passages, all alike'

Before you looms one of the most complex and utterly intimidating systems ever written. Linux, the free UNIX clone for the personal computer, produced by a mishmash team of UNIX gurus, hackers, and the occasional loon. The system itself reflects this complex heritage, and although the development of Linux may appear to be a disorganized volunteer effort, the system is powerful, fast, and free. It is a true 32-bit operating system solution."[1]

Much has happened in the Linux world in the three years since the first edition of this book was published. The 2.6 kernel brought many new innovations. Major players in the embedded software market, such as Wind River, have been forced to take notice and have jumped on the Linux bandwagon in one form or another. A plethora of consumer devices including cell phones, PDAs, and TV set-top boxes now run on Linux.

Even Microsoft has had to respond to the Open Source movement that spawned Linux. In the past couple of years, the Borg of Redmond has made available large chunks of its source code under a program called the "Shared Source initiative." This includes the entire source code base of Windows CE, the company's embedded, real-time operating system. You can bet that wouldn't have happened in the absence of Open Source.

---

[1] *Linux Installation and Getting Started*, Matt Welsh, et al.

I began the preface to the first edition by confessing that I've never really liked Unix, considering it deliberately obscure and difficult to use. Initially, Linux did little to change that impression and I still have something of a love/hate relationship with the thing.

But, while Linux is far from being ready for prime time in the world of desktop computing, there are some good things about it that have forced me to soften my bias and grin and bear it. In the embedded space where I work, Linux can no longer be ignored or avoided, nor should it be. A number of recent surveys have reported Linux capturing an increasing share of the embedded operating system marketplace.

Linux is indeed complex and, unless you're already a Unix guru, the learning curve is quite steep. The information is out there on the web but it is often neither easy to find nor readable. There are probably hundreds of books in print on Linux covering every aspect from beginners' guides to the internal workings of the kernel. But until recently little has been written about Linux in embedded or real-time environments.

I decided to climb the Linux learning curve partly because I saw it as an emerging market opportunity and partly because I was intrigued by the Open Source development model. The idea of programmers all over the world contributing to the development of a highly sophisticated operating system just for the fun of it is truly mind-boggling. Having the complete source code not only allows you to modify it to your heart's content, it allows you (in principle at least) to understand how the code works. Unfortunately, my experience has been that a lot of Linux code is "write-only." Someone obviously wrote it, but no one else can read it.

Open Source has the potential to be a major paradigm shift in how our society conducts business because it demonstrates that cooperation can be as useful in developing solutions to problems as competition. Yet at the time this book is being written, serious questions are being raised concerning whether or not it is possible to actually make money with Open Source software. Is there a business model that works? The jury is still out.

Audience and Prerequisites

My motivation for writing this book was to create the kind of book I wished I had had when I started out with Linux. With that in mind, the book is directed at two different audiences:

1. The primary audience is embedded programmers who need an introduction to Linux in the embedded space. This is where I came from and how I got into Linux so it seems like a reasonable way to structure the book.

2. The other audience is Linux programmers who need an introduction to the concepts of embedded and real-time programming.

Consequently, each group will likely see some material that is review although it may be presented with a fresh perspective.

This book is not intended as a beginners' guide. I assume that you have successfully installed a Linux system and have at least played around with it some. You know how to log in, you've experimented with some of the command utilities and have probably fired up a GUI desktop. Nonetheless, Chapter 2 is a cursory introduction to some of the features and characteristics of Linux that are of interest to embedded and real-time programmers.

The book is divided into two parts: Part I deals with Linux in the embedded space while Part II looks at different approaches to giving the Linux kernel deterministic real-time characteristics. It goes without saying that you can't learn to program by reading a book. You have to do it. That's why this book is designed as a practical hands-on guide. The companion CD contains several packages that we'll explore in depth in the following chapters.

Embedded programming generally implies a target machine that is separate and distinct from the workstation development machine. The target environment we'll be working with in this book is the x86. More and more embedded projects are choosing the PC architecture as a target platform because it's relatively inexpensive and there's lots of software available for it. And of course Linux was originally developed for the PC. For the purpose of this book, an x86 target need be nothing more than an old 486 or Pentium II box gathering dust in your closet. You don't even need a keyboard or monitor.

Personal Biases

Like most computer users, for better or worse, I've spent years in front of a Windows screen. But before that I was perfectly at home with DOS and even before that I hacked away at RT-11, RSX-11 and VMS. So it's not like I don't understand command line programming. In fact, back in the pre-Windows 95 era, it was probably a couple of years before I finally added WIN to my AUTOEXEC.BAT file.

Hardcore Unix programmers, on the other hand, think GUIs are for wimps. They proudly do everything from the command line. Say what you will, but I like GUIs. Yes, the command line still has its place, particularly for shell scripts and makefiles, but for moving around the file hierarchy and doing simple file operations like move, copy, delete, rename, etc, drag-and-drop beats the miserably obscure Unix commands hands down. I also refuse to touch text-based editors like vi and emacs. Sure they're powerful if you can remember all those obscure commands. Give me a WYSIWYG editor any day.

My favorite GUI is the KDE desktop environment. It has all the necessary bells and whistles including a very nice syntax coloring editor, not to mention a complete suite of office and personal productivity tools. KDE is included in most commercial Linux distributions. Clearly, you're free to use whatever environment you're most comfortable with to work the book's examples. But if you're new to Linux, I would recommend KDE.

OK, enough philosophizing. Let's get on with it. Join me for a thrill-packed, sometimes bumpy, but ultimately fun and rewarding, ride through those twisty little passages known as Linux.

Resources

www.Intellimetrix.us – The Downloads page of my website will host any updates to the book.

http://groups.yahoo.com/group/EmbeddedLinuxBook – This is a Yahoo Group discussion forum for the book.

About the Author

Doug Abbott is the principal of Intellimetrix, a consulting firm in Silver City, NM, specializing in hardware and software for industrial and scientific data acquisition and embedded product applications. Among his past and present clients are Agilent Technologies, Tektronix, Sandia National Laboratory and numerous smaller high-tech companies in Silicon Valley.

Mr. Abbott has over thirty years experience in various aspects of computer hardware and software design and has been an independent consultant for the past fifteen years. Prior to founding Intellimetrix, he managed software development for DSP Technology, Inc, a leading supplier of high-speed instrumentation.

Doug is also a popular instructor and seminar leader, who teaches classes in PC technology and real-time programming for University of California Berkeley Extension. He has taught the techniques of real-time programming and multi-tasking operating systems to hundreds of professional engineers. These classes can also be customized and made available for on-site presentation.

Mr. Abbott received an MSEE degree from the University of California at Berkeley.

What's on the CD-ROM?

README – this file
busybox-1.00.tar.bz2 – BusyBox
tinylogin-1.4.tar.bz2 – TinyLogin

/BlueCat

bc-5.0-x86-lite.tar.gz – BlueCat Linux version 5.0, including Linux kernel 2.6.0
BlueCatdemo.tar.gz – Source code for exercises in book
W95boot.img – DOS diskette image for testing a target computer
term.c – Source code for the serial port echo program on W95boot.img
rawrite.exe – Utility for writing W95boot.img to a diskette

/rtai

rtai-3.1.tar.bz2 – RTAI real-time extensions, version 3.1
Rtdemos.tar.gz – Source code for exercises in book
linux-2.6.7.tar – Linux kernel sources, version 2.6.7
showroom.tgz – Examples and demos for RTAI

/tools

eclipse-SDK-3.1.2-linux-gtk.tar.gz – Eclipse integrated development environment, version 3.1.2
gcc-2.95.3bin.tar.gz – GCC compiler, version 2.95.3, if needed
jre-1_5_0_06-linux-i586.bin – Java runtime environment, required by Eclipse
post-halloween-2.6.txt – A collection of notes, hints and "gotchas" about the 2.6 series kernel

/uClinux

m68k-elf-tools-incl-gdb.tar.gz – Cross-development toolchain for M68k processors
uClinux-2.4.x.tar.gz – 2.4 series Linux kernel patched for uClinux
uClinux-dist.tar.gz – "userland" sources, libraries, root filesystem, etc. for uCdimm module

Instruction for the installation and use of these files is found in the book.

Updates and corrections will be posted at either www.intellimetrix.us and/or http://groups.yahoo.com/group/embeddedlinuxbook.

CHAPTER 1

The Embedded and Real-Time Space

"Software is like sex, it's better when it's free."
—Linus Torvalds

What is Embedded?

You're at a party when an attractive member of the opposite sex approaches and asks you what you do. You could be flip and say something like "as little as possible," but eventually the conversation will get around to the fact that you write software for embedded systems. Before your new acquaintance starts scanning around the room for a lawyer or doctor to talk to, you'd better come up with a captivating explanation of embedded systems.

I usually start by saying that an embedded system is a device that has a computer inside it, but the user of the device doesn't necessarily know, or care, that the computer is there. It's hidden. The example I usually give is the engine control computer in your car. You don't drive the car any differently because the engine happens to be controlled by a computer. Oh, and there's a computer that controls the anti-lock brakes, another to decide when to deploy the airbags, and any number of additional computers that keep you entertained and informed as you sit in the morning's bumper-to-bumper traffic. In fact the typical car today has more raw computing power than the Lunar Lander.

You can then go on to point out that there are a lot more embedded computers out in the world than there are PCs. In fact, recent market data shows that PCs account for only about 2% of the microprocessor chips sold every year. The average house contains perhaps a couple dozen computers even if it doesn't have a PC.

From the viewpoint of programming, embedded systems show a number of significant differences from conventional "desktop" applications. For example, most desktop applications deal with a fairly predictable set of input/output (I/O) devices—a disk, graphic display, a keyboard, mouse, sound card, and network interface. And these devices are generally well supported by the operating system. The application programmer doesn't need to pay much attention to them.

Embedded systems on the other hand often incorporate a much wider variety of I/O devices than typical desktop computers. A typical system may include user I/O in the form of switches, pushbuttons, and various types of displays. It may have one or more communication channels, either asynchronous serial or network ports. It may implement data acquisition and control in the form of analog-to-digital (A/D) and digital-to-analog (D/A) converters. These devices seldom have the kind of operating system support that application programmers are accustomed to. Therefore the embedded-systems programmer often has to deal directly with the hardware.

What is Real-Time?

Real-time is even harder to explain. The basic idea behind real-time is that we expect the computer to respond to its environment *in time*. But what does "in time" mean? Many people assume that real-time means real fast. Not true. Real-time simply means *fast enough* in the context in which the system is operating. If we're talking about the computer that runs your car's engine, that's fast! That guy has to make decisions—about fuel flow, spark timing—every time the engine makes a revolution.

On the other hand, consider a chemical refinery controlled by one or more computers. The computer system is responsible for controlling the process and detecting potentially destructive malfunctions. But chemical processes have a time constant in the range of seconds to minutes at the very least. So we would assume that the computer system should be able to respond to any malfunction in sufficient time to avoid a catastrophe.

But suppose the computer were in the midst of printing an extensive report about last week's production or payroll when the malfunction occurred. How soon would it be able to respond to the potential emergency?

The essence of real-time computing is not only that the computer responds to its environment fast enough, but that it responds *reliably* fast enough. The engine control computer must be able to adjust fuel flow and spark timing every time the engine turns over. If it's late, the engine doesn't perform right. The controller of a chemical plant must be able to detect and respond to abnormal conditions in sufficient time to avoid a catastrophe. If it doesn't, it has failed.

So the art of real-time programming is designing systems that reliably meet timing constraints in the midst of random asynchronous events. Not surprisingly this is easier said than done and there is an extensive body of literature and development work devoted to the theory of real-time systems.

How and Why Does Linux Fit In?

By now just about everyone in the computer business knows the history of Linux; how Linus Torvalds started it all back in 1991 as a simple hobby project to which he invited other interested hackers to contribute. Back then no one could have predicted that this amorphous consortium of volunteer programmers and "the occasional loon," connected only by the Internet, would produce a credible operating system to compete with even the Borg of Redmond.

Of course, Linux developed as a general-purpose operating system in the model of Unix whose basic architecture it emulates. No one would suggest that Unix is suitable as an embedded or real-time operating system. It's big, it's a resource hog, and its scheduler is based on "fairness" rather than priority. In short, it's the exact antithesis of an embedded operating system.

But Linux has several things going for it that earlier versions of Unix lack. It's free and you get the source code. There is a large and enthusiastic community of Linux developers and users. There's a good chance that someone else either is working or has worked on the same problem you're facing. It's all out there on the web. The trick is finding it.

Open Source

Linux has been developed under the philosophy of Open Source software pioneered by the Free Software Foundation. Quite simply, Open Source is based on the notion that software should be freely available: to use, to modify, and to copy. The idea has been around for some 20 years in the technical culture that built the Internet and the World Wide Web and in recent years has spread to the commercial world.

There are a number of misconceptions about the nature of Open Source software. Perhaps the best way to explain what it is, is to start by talking about what it isn't.

- *Open Source is not shareware.* A precondition for the use of shareware is that you pay the copyright holder a fee. Open Source code is freely available and there is no obligation to pay for it.

- *Open Source code is not public domain.* Public domain code, by definition, is not copyrighted. Open Source code is copyrighted by its author who has released it under the terms of an Open Source software license. The copyright owner thus gives you the right to use the code provided you adhere to the terms of the license.
- *Open Source is not necessarily free of charge.* Having said that there's no obligation to pay for Open Source software doesn't preclude you from charging a fee to package and distribute it. A number of companies are in the specific business of selling packaged "distributions" of Linux.

Why would you pay someone for something you can get for free? Presumably because everything is in one place and you can get some support from the vendor. Of course the quality of support greatly depends on the vendor.

So "free" refers to freedom to use the code and not necessarily zero cost. Think "free speech," not "free beer."

Open Source code is:

- subject to the terms of an Open Source license, in many cases the GNU Public License (see below);
- subject to critical peer review. As an Open Source programmer, your code is out there for everyone to see and the Open Source community tends to be a very critical group. Open Source code is subject to extensive testing and peer review. It's a Darwinian process in which only the best code survives. "Best" of course is a subjective term. It may be the best *technical* solution but it may also be completely unreadable;
- highly subversive—the Open Source movement subverts the dominant paradigm, which says that intellectual property such as software must be jealously guarded so you can make a lot of money off of it. In contrast, the Open Source philosophy is that software should be freely available to everyone for the maximum benefit of society. Richard Stallman, founder of the Free Software Foundation, is particularly vocal in advocating that software should not have owners (see Appendix C).

Not surprisingly, Microsoft sees Open Source as a serious threat to its business model. Microsoft representatives have gone so far as to characterize Open Source as "un-American." On the other hand, many leading vendors of Open Source software give their programmers and engineers company time to contribute to the Open Source community. And it's not just charity, it's good business!

Portable and Scalable

Linux was originally developed for the Intel x86 family of processors and most of the ongoing kernel development work continues to be on x86s. Nevertheless, the design of the Linux kernel makes a clear distinction between processor-dependent code, which must be modified for each different architecture, and code that can be ported to a new processor simply by recompiling it. Consequently, Linux has been ported to a wide range of 32-bit, and more recently 64-bit, processor architectures including:

- Motorola 68k and its many variants
- Alpha
- Power PC
- ARM
- Sparc
- MIPS

to name a few of the more popular. So whatever 32-bit architecture you're considering for your embedded project, chances are there's a Linux port available for it and a community of developers supporting it.

A typical desktop Linux installation runs into a couple of gigabytes of disk space and requires 256 Mbytes or more of random access memory (RAM) to execute decently. By contrast, embedded targets are often limited to 64 Mbytes or less of RAM and perhaps 128 Mbytes of flash read-only memory (ROM) for storage. Fortunately, Linux is highly modular. Much of that couple of gigabytes represents documentation, desktop utilities and options like games that simply aren't necessary in an embedded target. It is not difficult to produce a fully functional, if limited, Linux system occupying no more than 2 Mbytes of flash memory.

The kernel itself is highly configurable and includes reasonably user-friendly tools that allow you to remove kernel functionality not required in your application.

Where is Linux Embedded?

Just about everywhere. As of July 2005, the website *LinuxDevices.com* listed a little over 300 commercially available products running Linux. They range from cell phones, personal digital assistants (PDAs) and other handheld devices through routers and gateways, thin clients, multimedia devices, and TV set-top boxes to robots and even rugged VME chassis suitable for military command and control applications. And these are just the products the *LinuxDevices* editors happen to know about.

One of the first, and perhaps best known home entertainment devices to embed Linux is the TiVo personal video recorder (PVR) that created a revolution in television viewing when it was first introduced in 2000. The TiVo is based on a Power PC processor and runs a "home grown" embedded Linux port that uses a graphics rendering chip for generating video.

Half the fun of having a device that runs Linux is making it do something more, or different, than the original manufacturer intended. There are a number of websites and books devoted to hacking the TiVo. Increasing the storage capacity is perhaps the most obvious hack. Other popular hacks include displaying weather, sports scores, or stock quotes, and setting up a web server.

Applications for embedded Linux aren't limited to consumer products. It's found in point-of-sale terminals, video surveillance systems, robots, even in outer space. NASA's Goddard Space Flight Center developed a version of Linux called *FlightLinux* to address the unique problems of spacecraft onboard computers. On the International Space Station, Linux-based devices control the rendezvous and docking operations for unmanned servicing spacecraft called *Automatic Transfer Vehicles*.

Historically, telecommunications carriers and service providers have relied on specialized, proprietary platforms to meet the availability, reliability, performance, and service response time requirements of telecommunication networks. Today, carriers and service providers are embracing "open architecture" and moving to COTS (commercial off-the-shelf) hardware and software in an effort to drive down costs while still maintaining carrier class performance.

Linux plays a major role in the move to an open, standards-based network infrastructure. In 2002, the Open Source Development Lab (OSDL) set up a working group to define "carrier grade Linux" (CGL) in an effort to meet the higher availability, serviceability, and scalability requirements of the telecom industry. The objective of CGL is to achieve a level of reliability known as *five nines*, meaning the system is operational 99.999% of the time. That translates into no more than about 5 minutes of downtime in a year.

Open Source Licensing

Most end-user license agreements (EULA) for software are specifically designed to restrict what you are allowed to do with the software covered by the license. Typical restrictions prevent you from making copies or otherwise redistributing it. You are often admonished not to attempt to "reverse-engineer" the software.

By contrast, an Open Source license is intended to guarantee your rights to use, modify, and copy the subject software as much as you'd like. Along with the rights comes an obligation. If you modify and subsequently distribute software covered by an Open Source license, you are obligated to make available the modified source code under the same terms. The changes become a "derivative work" which is also subject to the terms of the license. This allows other users to understand the software better and to make further changes if they wish.

Arguably the best-known, and most widely used, Open Source license is the general public license (GPL) first released by the Free Software Foundation (FSF) in 1989. The Linux kernel is licensed under the GPL, but the GPL has a problem that makes it unworkable in many commercial situations. Software that does nothing more than *link* to a library released under the GPL is considered a derivative work and is therefore subject to the terms of the GPL and must be made available in source code form.

To get around this, and thus promote the development of Open Source libraries, the FSF came up with the "library GPL (LGPL)." The distinction is that a program linked to a library covered by the LGPL is not considered a derivative work and so there's no requirement to distribute the source, although you must still make available the source to the library itself.

Subsequently, the LGPL became known as the *lesser GPL* because it offers less freedom to the user. So while the LGPL makes it possible to develop proprietary products using Open Source software, the FSF encourages developers to place their libraries under the GPL in the interest of maximizing openness.

At the other end of the scale is the Berkeley software distribution (BSD) license, which predates the GPL by some 12 years. It "suggests," but does not require, that source code modifications be returned to the developer community and it specifically allows derived products to use other licenses, including proprietary ones.

Other licenses—and there are quite a few—fall somewhere between these two poles. The Mozilla public license (MPL) for example, developed in 1998 when Netscape made its browser open-source, contains more requirements for derivative works than the BSD license, but fewer than the GPL or LGPL. The Open Source Initiative (OSI), a nonprofit group that certifies licenses meeting its definition of Open Source, lists 58 certified licenses on its website as of September 2005.

At the time this is being written, the FSF has begun a major effort to revise and update the GPL, which was last updated in 1991. The goal is to clarify the language and, in the spirit of internationalization, make it easier to translate. Issues surrounding software patents are also expected to be addressed. The scope of the project is such that an initial draft is expected in early 2006 with the final release of the version 3 GPL coming sometime in 2007.

Legal Issues

Considerable FUD[1] has been generated about the legal implications of Open Source, particularly in light of SCO's claims that the Linux kernel is "contaminated" with its proprietary Unix code. The SCO Group, formerly known as *Santa Cruz Operations*, acquired the rights to the Unix System V source code from Novell in 1996, although there is some dispute as to exactly what SCO bought from Novell. In any case, SCO asserts that IBM introduced pieces of SCO's copyrighted, proprietary Unix code into the Linux kernel and is demanding license fees from Linux users as a consequence.

So all of a sudden there is serious money to be made by fighting over Open Source licensing issues. The upshot is that embedded developers need to be aware of license issues surrounding both Open Source and proprietary software. Of necessity, embedded software is often intimately tied to the operating system and includes elements derived or acquired from other sources. While no one expects embedded-systems engineers to be intellectual property attorneys, it is nevertheless essential to understand the license terms of the software you use and create to be sure that all the elements "play nicely" together.

And the issue cuts both ways. There are also efforts to identify violations of the GPL. The intent here is not to make money, but to defend the integrity of the GPL by putting pressure on violators to clean up their act. In particular, the GPL Violations Project has "outed" a dozen or so embedded Linux vendors who appear to have played fast and loose with the GPL terms. According to Harald Welte, founder of the GPL Violations Project, the most common offenders are networking devices such as routers, followed by set-top boxes and vehicle navigation systems.

Open source licensing expert Bruce Perens has observed that embedded developers seem to have a mindset that "this is embedded, no one can change the source—so the GPL must not really apply to us." It does.

[1] Fear, uncertainty and doubt.

Resources

Linux resources on the web are extensive. This is a list of some sites that are of particular interest to embedded developers.

www.embedded.com – The website for *Embedded Systems Programming* magazine. This site is not specifically oriented to Linux, but is quite useful as a more general information tool for embedded system issues.

www.fsf.org – The Free Software Foundation.

www.keegan.org/jeff/tivo/hackingtivo.html – One of many sites devoted to TiVo hacks.

www.kernel.org – The Linux kernel archive. This is where you can download the latest kernel versions as well as virtually any previous version.

www.linuxdevices.com – A news and portal site devoted to the entire range of issues surrounding embedded Linux.

www.opensource.org – The Open Source Initiative (OSI), a nonprofit corporation "dedicated to managing and promoting the Open Source Definition for the good of the community." OSI certifies software licenses that meet its definition of Open Source.

www.osdl.org – Open Source Development Lab, a nonprofit consortium focused on accelerating the growth and adoption of Linux in both the enterprise and, more recently, embedded spaces. In September of 2005, OSDL took over the work of the Embedded Linux Consortium, which had developed a "platform specification" for embedded Linux.

www.gpl-violations.org – The GPL Violations Project was started to "raise the awareness about past and present violations" of the GPL. According the website it is "still almost a one-man effort."

www.sourceforge.net – "World's largest Open Source development website." Provides free services to open source developers including project hosting and management, version control, bug and issue tracking, backups and archives, and communication and collaboration resources.

www.uclinux.org – The Linux/Microcontroller project is a port of Linux to systems without a Memory Management Unit (MMU).

www.slashdot.org – "News for nerds, stuff that matters." A very popular news and forum site focused on open source software in general and Linux in particular.

CHAPTER 2

Introducing Linux

"There are two major products to come out of Berkeley: LSD and Unix. We don't believe this to be a coincidence."

—Anonymous

For those who may be new to Unix-style operating systems, this chapter provides an introduction to some of the salient features of Linux, especially those of interest to embedded-system developers. This is by no means a thorough introduction and there are many books available that delve into these topics in much greater detail.

Feel free to skim, or skip this chapter entirely, if you are already comfortable with Unix and Linux concepts.

Features

Here are some of the important features of Linux and Unix-style operating systems in general:

- *Multitasking*. The Linux scheduler implements true, preemptive multitasking in the sense that a higher priority process made ready by the occurrence of an asynchronous event will preempt the currently running process. However, the stock Linux kernel itself is not preemptible[1]. So a process may not be preempted while it is executing a kernel service. Some kernel services can be rather long and the resulting latencies make standard Linux generally unsuitable for real-time applications.
- *Multiuser*. Unix evolved as a time-sharing system that allowed multiple users to share an expensive (at that time anyway) computer. Thus there are a num-

[1] Is it "preemptible" or "preemptable"? Word 2000's spell checker says they're both wrong. A debate on *linuxdevices.com* a while back seemed to come down on the side of "ible" but not conclusively. I think I'll stick with preemptible.

ber of features that support privacy and data protection. Linux preserves this heritage and puts it to good use in server environments.[2]

- *Multiprocessing*. Linux offers extensive support for true symmetric multiprocessing (SMP) where multiple processors are tightly coupled through a shared memory bus.
- *Protected Memory*. Each Linux process operates in its own private memory space and is not allowed to directly access the memory space of another process. This prevents a wild pointer in one process from damaging the memory space of another process. The errant access is trapped by the processor's memory protection hardware and the process is terminated with appropriate notification.
- *Hierarchical File System*. Yes, all modern operating systems—even DOS—have hierarchical file systems. But the Linux/Unix model adds a couple nice wrinkles on top of what we're used to with traditional PC operating systems:
  - *Links*. A link is simply a file system entry that points to another file rather than being a file itself. Links can be a useful way to share files among multiple users and find extensive use in configuration scenarios for selecting one of several optional files.
  - *Device-Independent I/O*. Again, this is nothing new, but Linux takes the concept to its logical conclusion by treating every peripheral device as an entry in the file system. From an application's viewpoint, there is absolutely no difference between writing to a file and writing to, say, a printer.

Protected Mode Architecture

The implementation of protected mode memory in contemporary Intel processors first made its appearance in the 80386. It utilizes a full 32-bit address for an addressable range of 4 Gbytes. Access is controlled such that a block of memory may be: executable, read only, or read/write.

The processor can operate in one of four "privilege levels." A program running at the highest privilege level, level 0, can do anything it wants—execute I/O instructions, enable and disable interrupts, modify descriptor tables. Lower privilege levels prevent programs from performing operations that might be "dangerous." A word processing application probably shouldn't be messing with interrupt flags, for example. That's the job of the operating system.

[2] Although my experience in the embedded space is that the protection features, particularly file permissions, can be a downright nuisance. Some programs, the Firefox browser is an example, don't have the courtesy to tell you that you don't have permission to write the file, they just sit there and do nothing.

So application code typically runs at the lowest level while the operating system runs at the highest level. Device drivers and other services may run at the intermediate levels. In practice however, Linux and most other operating systems for Intel processors only use levels 0 and 3. In Linux level 0 is called *kernel space* while level 3 is called *user space*.

Real Mode

To begin our discussion of protected mode programming in the x86, it's useful to review how "real" address mode works.

Back in the late 1970s when Intel was designing the 8086, the designers faced the dilemma of how to access a megabyte of address space with only 16 bits. At the time a megabyte was considered an immense amount of memory. The solution they came up with, for better or worse, builds a 20-bit (1 megabyte) address out of two 16-bit quantities called the *segment* and *offset*. Shifting the segment value four bits to the left and adding it to the offset creates the 20-bit linear address (see Figure 2-1).

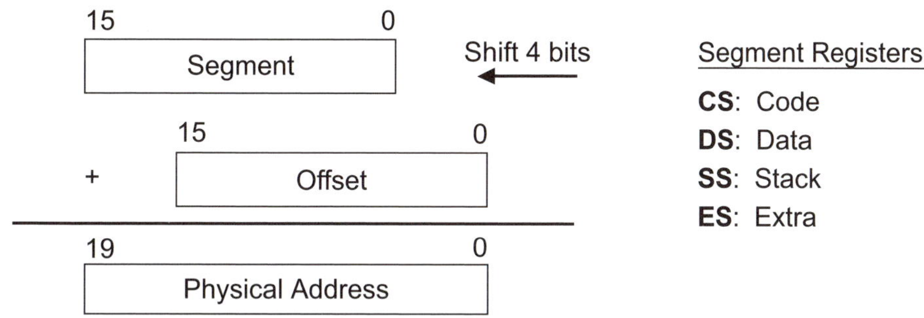

Figure 2-1: X86 Real Mode Addressing

x86 processors have four segment registers in real mode. Every reference to memory derives its segment value from one of these registers. By default, instruction execution is relative to the code segment (CS), most data references, the (MOV) instruction for example, are relative to the data segment (DS), and instructions that reference the stack are relative to the stack segment (SS). The extra segment (ES) is used in string move instructions and can be used whenever an extra data segment is needed. The default segment selection can be overridden with segment prefix instructions.

A segment can be up to 64 Kbytes long and is aligned on 16-byte boundaries. Programs less than 64 Kbytes are inherently position-independent and can be easily relocated anywhere in the 1-Mbyte address space. Programs larger than 64 Kbytes, either in code or data, require multiple segments and must explicitly manipulate the segment registers.

Protected Mode

Protected mode still makes use of the segment registers, but instead of providing a piece of the address directly, the value in the segment register (now called the *selector*) becomes an index into a table of *segment descriptors*. The segment descriptor fully describes a block of memory including, among other things, its base and limit (see Figure 2-2). The linear address in physical memory is computed by adding the offset in the logical address to the base contained in the descriptor. If the resulting address is greater than the limit specified in the descriptor, the processor signals a memory protection fault.

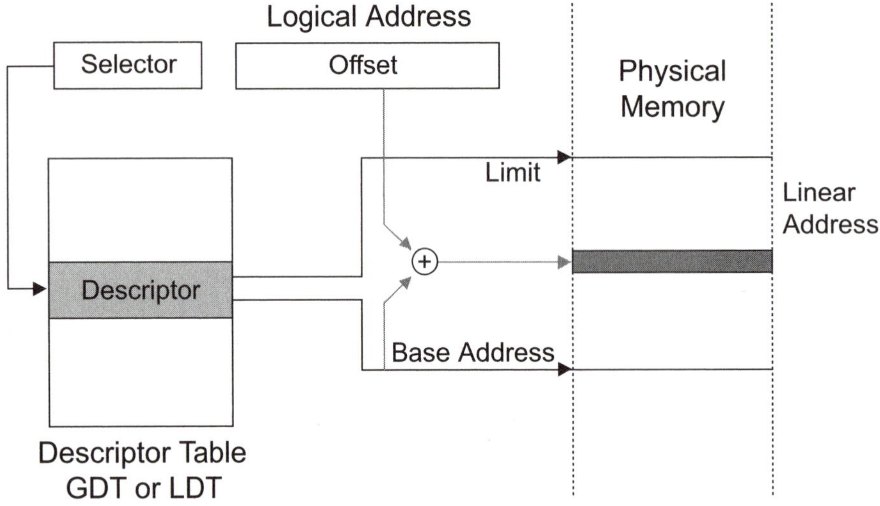

Figure 2-2: Protected Mode Address Calculation

A descriptor is an 8-byte object that tells us everything we need to know about a block of memory.

Base Address[31:0] Starting address for this block/segment.

Limit[19:0] *Length of this segment*. This may be either the length in bytes (up to 1 Mbyte) or the length in 4-Kbyte *pages*. The interpretation is defined by the Granularity bit.

Type A 4-bit field that defines the kind of memory that this segment describes

S 0 = This descriptor describes a "System" segment. 1 = This descriptor describes a code or data segment.

DPL	*Descriptor Privilege Level.* A 2-bit field that defines the minimum privilege level required to access this segment.
P	*Present.* 1 = The block of memory represented by this descriptor is present in memory. Used in paging.
G	*Granularity.* 0 = Interpret Limit as bytes. 1 = Interpret Limit as 4 Kbyte pages

Note that with the Granularity bit set to 1, a single segment descriptor can represent the entire 4-Gbyte address space.

Normal descriptors (S bit = 1) describe memory blocks representing data or code. The Type field is four bits where the most significant bit distinguishes between code and data segments. Code segments are executable, data segments are not. A code segment may or may not also be readable. A data segment may be writable. Any attempted access that falls outside the scope of the Type field—attempting to execute a data segment for example—causes a memory protection fault.

"Flat" vs. Segmented Memory Models

Because a single descriptor can reference the full 4-Gbyte address space, it is possible to build your system by reference to a single descriptor. This is known as *flat* model addressing and is, in effect, a 32-bit equivalent of the addressing model found in most 8-bit microcontrollers as well as the "tiny" memory model of DOS. All memory is equally accessible and there is no protection.

In order to take advantage of the protection features built into Protected Mode, you must allocate different descriptors for the operating system and applications. Figure 2-3 illustrates the difference between flat model and segmented model.

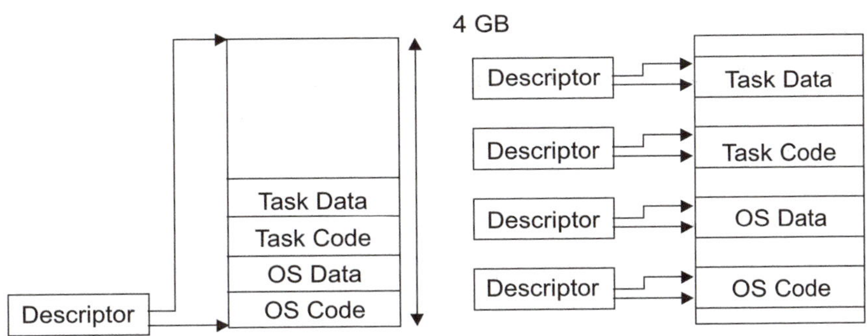

Figure 2-3: "Flat" vs. Segmented Addressing

Paging

Paging is the mechanism that allows each task to pretend that it owns a very large flat address space. That space is then broken down into 4-Kbyte *pages*. Only the pages currently being accessed are kept in main memory. The others reside on disk.

As shown in Figure 2-4, paging adds another level of indirection. The 32-bit linear address derived from the selector and offset is divided into three fields. The high order 10 bits serve as an index into the *page directory*. The page directory entry (PDE) points to a *page table*. The next ten bits in the linear address provide an index into that table. The page table entry (PTE) provides the base address of a 4-Kbyte page in physical memory called a *page frame*. The low order 12 bits of the original linear address supplies the offset into the page frame. Each task has its own page directory pointed to by processor control register CR3.

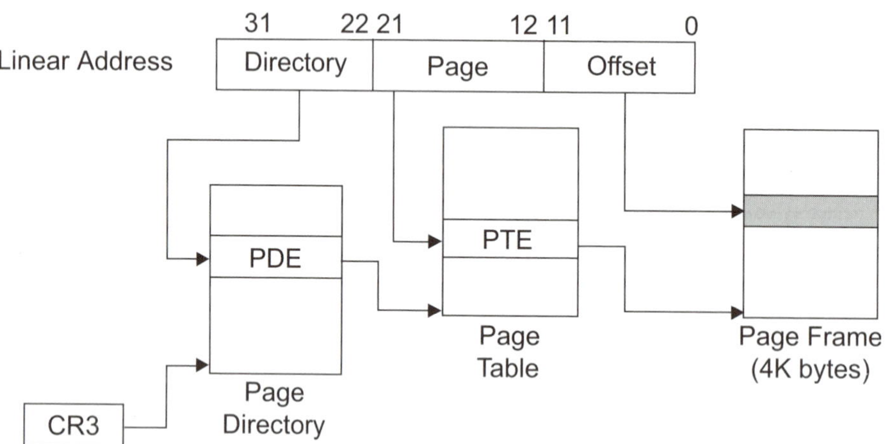

Figure 2-4: Paging

At either stage of this lookup process it may turn out that either the page table or the page frame is not present in physical memory. This causes a *page fault*, which in turn causes the operating system to find the corresponding page on disk and load it into an available page in memory. This in turn may require "swapping out" the page that currently occupies that memory.

A further advantage to paging is that it allows multiple tasks or processes to easily share code and data by simply mapping the appropriate sections of their individual address spaces into the same physical pages.

Paging is optional, you don't have to use it, although Linux does. Paging is controlled by a bit in processor register CR0.

Page directory and page table entries are each 4 bytes long, so the page directory and page tables are a maximum of 4 Kbytes, which also happens to be the page frame size. The high-order 20 bits point to the base of a page table or page frame. Bits 9 to 11 are available to the operating system for its own use. Among other things, these could be used to indicate that a page is to be "locked" in memory, i.e., not swappable.

Of the remaining control bits the most interesting are:

P *Present*. 1 = this page is in memory. If this bit is 0, referencing this page directory or page table entry causes a page fault. Note that when P == 0 the remainder of the entry is not relevant.

A *Accessed*. 1 = this page has been read or written. Set by the processor but cleared by the operating system (OS). By periodically clearing the accessed bits, the OS can determine which pages haven't been referenced in a long time and are therefore subject to being swapped out.

D *Dirty*. 1 = this page has been written. Set by the processor but cleared by the OS. If a page has not been written to, there is no need to write it back to disk when it has to be swapped out.

The Linux Process Model

The basic structural element in Linux is a *process* consisting of executable code and a collection of *resources* like data, file descriptors, and so on. These resources are fully protected such that one process can't directly access the resources of another. In order for two processes to communicate with each other, they must use the inter- process communication mechanisms defined by Linux such as shared memory regions or pipes.

This is all well and good as it establishes a high degree of protection in the system. An errant process will most likely be detected by the system and thrown out before it can do any damage to other processes (see Figure 2-5). But there's a price to be paid in terms of excessive overhead in creating processes and using the inter-process communication mechanisms.

A *thread* on the other hand is code only. Threads only exist within the context of a process and all threads in one process share its resources. Thus all threads have equal access to data memory and file descriptors. This model is sometimes called *lightweight multitasking* to distinguish it from the Unix/Linux process model.

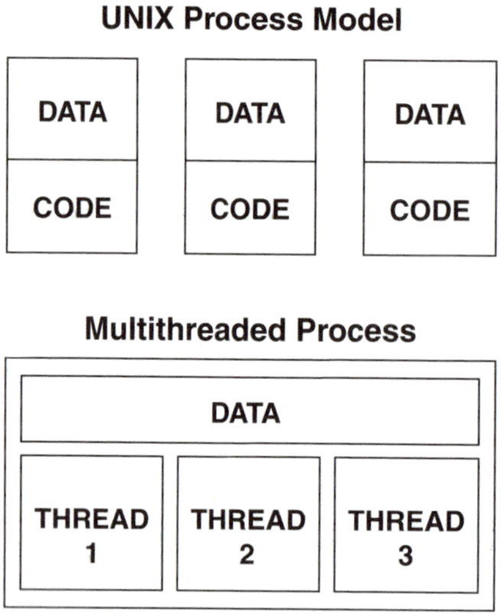

UNIX Process Model

DATA	DATA	DATA
CODE	CODE	CODE

Multithreaded Process

DATA		
THREAD 1	THREAD 2	THREAD 3

Figure 2-5: "Processes" vs. "Threads"

The advantage of lightweight tasking is that inter-thread communication is more efficient. The drawback of course is that any thread can clobber any other thread's data. Historically, most real-time operating systems have been structured around the lightweight model. In recent years of course the cost of memory protection hardware has dropped dramatically. In response, many real-time operating system (RTOS) vendors now offer protected mode versions of their systems that look like the Linux process model.

The fork() Function

Linux starts life with one process, the init process, created at boot time. Every other process in the system is created by invoking `fork()`. The process calling `fork()` is termed the *parent* and the newly created process is termed the *child*. So every process has ancestors and may have descendants depending on who created who.

If you've grown up with multitasking operating systems where tasks are created from functions by calling a task creation service, the fork process can seem downright bizarre. `fork()` creates a *copy of the parent process*—code, data, file descriptors, and any other resources the parent may currently hold. This could add up to megabytes of memory space to be copied. To avoid copying a lot of stuff that may be overwritten anyway, Linux employs a *copy-on-write strategy*.

fork() begins by making a copy of the process data structure and giving it a new process identifier (PID) for the child process. Then it makes a new copy of the page directory and page tables. Initially the page table entries all point to the same physical pages as the parent process. All pages for both processes are set to read-only. When one of the processes tries to write, that causes a page fault, which in turn causes Linux to allocate a new page for that process and copy over the contents of the existing page.

```
#include <unistd.h>
#include <>
pid_t pid;
void do_child_thing (void)
{
    printf ("I am the child. My PID is %d\n", pid);
}
void do_parent_thing (void)
{
    printf ("I am the parent. My child's PID is %d\n", pid);
}
void main (void)
{
    switch (pid = fork())
    {
            case -1:
                printf ("fork failed\n");
                break;
            case 0:
                do_child_thing();
                break;
            default:
                do_parent_thing();
        }
        exit (0);
}
```

Listing 2-1: Trivial Example of Fork

Since both processes are executing the same code, they both continue from the return from fork() (this is what's so bizarre!). In order to distinguish parent from child, fork() returns a function value of 0 to the child process but returns the PID of the child to the parent process. Listing 2-1 is a trivial example of the fork call.

clone() is a Linux-specific variation on fork(), the difference being that the former offers greater flexibility in specifying how much of the parent's operating environment is shared with the child. This is the mechanism that is used to implement Posix threads at the kernel level.

The execve() Function

Of course, what really happens 99% of the time is that the child process invokes a new program by calling execve() to load an executable image file from disk. Listing 2-2 shows in skeletal form a simple command line interpreter. It reads a line of text from stdin, parses it, and calls fork() to create a new process. The child then calls execve() to load a file and execute the command just entered. execve() overwrites the calling process's code, data, and stack segments.

If this is a normal "foreground" command, the command interpreter must wait until the command completes. This is accomplished with waitpid() which blocks the calling process until the process matching the pid argument has completed. Note by the way that most multitasking operating systems do not have the ability to block one process or task pending the completion of another.

If execve() succeeds, it does not return. Instead, control is transferred to the newly loaded program.

```
#include <unistd.h>
void main (void)
{
    char *argv[10], *filename;
    char text[80];
    char foreground;
    pid_t pid;
    int status;

    while (1)
    {
```

Listing 2-2: Command Line Interpreter *(continued)*

```
        gets (text);
// Parse the command line to derive filename and
// arguments. . Decide if it's a foreground command.
        switch (pid = fork())
        {
            case -1:
                printf ("fork failed\n");
                break;
            case 0:     // child process
                if (execve (filename, argv, NULL) < 0)
                    printf ("command failed\n");
                break;
            default:   // parent process
                if (foreground)
                    waitpid (&status, pid);
        }
    }
}
```

Listing 2-2: Command Line Interpreter

The Linux Filesystem

The Linux filesystem is in many ways similar to the filesystem you might find on a Windows PC or a Macintosh. It's a hierarchical system that lets you create any number of subdirectories under a root directory identified by "/". Like Windows, file names can be very long. However in Linux, as in most Unix-like systems, filename "extensions," the part of the filename following ".", have much less meaning. For example, while Windows executables always have the extension ".exe", Linux executables rarely have an extension at all. By and large, the contents of a file are identified by a file header rather than a specific extension identifier. Nevertheless, many applications, the C compiler for example, do support default file extensions.

Unlike Windows, file names in Linux are *case-sensitive*. Foobar is a different file from foobar which is different from fooBar. Sorting is also case-sensitive. File names beginning with upper case letters appear before those that begin with lower case letters in directory listings sorted by name. File names that begin with "." are considered to be "hidden" and are not displayed in directory listings unless you specifically ask for them.

Additionally, the Linux filesystem has a number of features that go beyond what you find in a typical Windows system. Let's take a look at some of the features that may be of interest to embedded programmers.

File Permissions

Because Linux is multiuser, every file has a set of "permissions" associated with it to specify what various classes of users are allowed to do with that file. Get a detailed listing of some Linux directory, either by entering the command ls –l in a console window or with the desktop file manager. Part of the entry for each file is a set of 10 flags and a pair of names that look something like this:

```
-rw-r--r-- Andy physics
```

In this example, Andy is the "owner" of the file and he belongs to a "group" of users called *physics*, perhaps the physics department at some university. Generally, but not always, the owner is the person who created the file.

The first of the 10 flags identifies the file type. Ordinary files get a dash here. Directories are identified by "d", links are "l" and so on. The remaining nine flags divide into three groups of three flags each. The flags are the same for all groups and represent, respectively, permission to read the file, "r", write the file, "w", or execute the file if it's an executable, "x". Write permission also allows the file to be deleted.

The three groups then represent the permissions granted to different classes of users. The first group identifies the permissions granted the owner of the file and virtually always allows reading and writing. The second flag group gives permissions to other members of the same group of users. In this case the physics group has read access to the file but not write access. The final flag group gives permissions to the "world," i.e., all users.

The "x" permission is worth a second look. In Windows, a file with the extension .exe is *assumed* to be executable. In Linux, a binary executable is identified by the "x" permission since we don't have an explicit file extension to identify it. Furthermore, only those classes of users whose "x" permission is set are allowed to invoke execution of the file. So if I'm logged in as an ordinary user, I'm not allowed to invoke programs that might change the state of the overall system such as changing network configuration, or installing or removing device drivers.

Another interesting feature of "x" is that it also applies to shell scripts, which we'll come to later in this chapter. For you DOS fans, a shell script is the same thing as a

.bat file. It's a text file of commands to be executed as a program. But the shell won't execute the script unless its "x" bit is set.

The "root" User

There's one very special user, named "root," in every Linux system. Root can do anything to any file regardless of the permission flags. Root is primarily intended for system administration purposes and is not recommended for day-to-day use. Clearly you can get in a lot of trouble if you're not careful and root privileges pose a potential security threat. Nevertheless, the kinds of things that embedded and real-time developers do with the system often require write or executable access to files owned by root and thus require you to be logged in as the root user.

If you're logged on as a normal user, you can switch to being root with the su, substitute user, command. The su command with no arguments starts up a shell with root privileges provided you enter the correct password. To return to normal user status, terminate the shell by typing ^d or exit.

In the past, I would just log in as root most of the time because it was less hassle. One consequence of this is that every file I created was owned by root and couldn't be written by an ordinary user without changing the permissions. It became a vicious circle. The more I logged in as root, the more I *had* to log in as root to do anything useful. I've since adopted the more prudent practice of logging in as a normal user and only switching to root when necessary.

The /proc Filesystem

The /proc filesystem is an interesting feature of Linux. It acts just like an ordinary file system. You can list the files in the /proc directory, you can read and write the files, but they don't really exist. The information in a /proc file is generated on the fly when the file is read. The kernel module that registered a given /proc file contains the functions that generate read data and accept write data.

/proc files are another window into the kernel. They provide dynamic information about the state of the system in a way that is easily accessible to user-level tasks and the shell. In the abbreviated directory listing of Figure 2-6, the directories with number labels represent processes. Each process gets a directory under /proc with several files describing the state of the process.

Try it out

Its interesting to see how many processes Linux spawns just by booting up. Reboot your system, log in and execute:

ps –A | more

PID	TTY	TIME	CMD
1	?	00:00:03	init
2	?	00:00:00	keventd
3	?	00:00:00	kapmd
4	?	00:00:00	ksoftirqd_CPU0
9	?	00:00:00	bdflush
5	?	00:00:00	kswapd
6	?	00:00:00	kscand/DMA
7	?	00:00:00	kscand/Normal
8	?	00:00:00	kscand/HighMem
10	?	00:00:00	kupdated
11	?	00:00:00	mdrecoveryd
15	?	00:00:00	kjournald
73	?	00:00:00	khubd
1180	?	00:00:00	kjournald
1489	?	00:00:00	syslogd
1493	?	00:00:00	klogd
1511	?	00:00:00	portmap
1531	?	00:00:00	rpc.statd
1598	?	00:00:00	apmd
1635	?	00:00:00	sshd
1649	?	00:00:00	xinetd
1666	?	00:00:00	ntpd
1686	?	00:00:00	sendmail
1695	?	00:00:00	sendmail
1705	?	00:00:00	gpm
1714	?	00:00:00	crond
1725	?	00:00:00	cupsd
1785	?	00:00:01	xfs
1794	?	00:00:00	anacron
1803	?	00:00:00	atd
1811	tty1	00:00:00	mingetty
1812	tty2	00:00:00	mingetty
1813	tty3	00:00:00	mingetty
1814	tty4	00:00:00	mingetty
1815	tty5	00:00:00	mingetty
1816	tty6	00:00:00	mingetty
1817	?	00:00:00	gdm-binary
1860	?	00:00:01	gdm-binary
1861	?	00:00:07	X
1870	?	00:00:00	startkde
1915	?	00:00:00	ssh-agent
1959	?	00:00:00	kdeinit
1962	?	00:00:00	kdeinit
1965	?	00:00:00	kdeinit
1967	?	00:00:10	kdeinit
1977	?	00:00:06	artsd
1990	?	00:00:04	kdeinit
1991	?	00:00:00	kwrapper
1993	?	00:00:04	kdeinit
1994	?	00:00:05	kdeinit
1996	?	00:00:05	kdeinit
1998	?	00:00:04	kdeinit
1999	?	00:00:00	kdeinit
2000	?	00:00:00	autorun
2006	?	00:00:03	kdeinit
2007	?	00:00:00	pam-panel-icon
2008	?	00:00:00	pam_timestamp_c
2009	?	00:00:00	eggcups
2010	?	00:00:03	rhn-applet-gui
2012	?	00:00:00	gconfd-2
2013	?	00:00:04	kdeinit
2017	pts/1	00:00:00	bash
2044	pts/1	00:00:00	ps

This comes from Red Hat 9.0 running the KDE desktop environment.

```
ls -l /proc
total 0
dr-xr-xr-x   3 root     root      0 Aug 25 15:23 1
dr-xr-xr-x   3 root     root      0 Aug 25 15:23 2
dr-xr-xr-x   3 root     root      0 Aug 25 15:23 3
dr-xr-xr-x   3 bin      root      0 Aug 25 15:23 303
dr-xr-xr-x   3 nobody   nobody    0 Aug 25 15:23 416
dr-xr-xr-x   3 daemon   daemon    0 Aug 25 15:23 434
dr-xr-xr-x   3 xfs      xfs       0 Aug 25 15:23 636
dr-xr-xr-x   4 root     root      0 Aug 25 15:23 bus
-r--r--r--   1 root     root      0 Aug 25 15:23 cmdline
-r--r--r--   1 root     root      0 Aug 25 15:23 cpuinfo
-r--r--r--   1 root     root      0 Aug 25 15:23 devices
-r--r--r--   1 root     root      0 Aug 25 15:23 filesystems
dr-xr-xr-x   2 root     root      0 Aug 25 15:23 fs
dr-xr-xr-x   4 root     root      0 Aug 25 15:23 ide
-r--r--r--   1 root     root      0 Aug 25 15:23 interrupts
-r--r--r--   1 root     root      0 Aug 25 15:23 ioports
```

Figure 2-6: The /proc Filesystem

Try it out

```
[root@lab /Doug]# cd /proc
[root@lab /proc]# cat interrupts
       CPU0
  0:   555681    XT-PIC  timer
  1:      110    XT-PIC  keyboard
  2:        0    XT-PIC  cascade
  7:        1    XT-PIC  soundblaster
  8:        1    XT-PIC  rtc
  9:     1670    XT-PIC  DC21041 (eth0)
 12:    11342    XT-PIC  PS/2 Mouse
 13:        1    XT-PIC  fpu
 14:   203260    XT-PIC  ide0
NMI:        0
```

The interrupts proc file tells you what interrupt sources have been registered by what device drivers and how many times each interrupt has triggered. To prove that the file data is being created dynamically, repeat the same command.

(continued)

```
[root@lab /proc]# cat interrupts
        CPU0
  0:   556540      XT-PIC  timer
  1:      116      XT-PIC  keyboard
  2:        0      XT-PIC  cascade
  7:        1      XT-PIC  soundblaster
  8:        1      XT-PIC  rtc
  9:     1672      XT-PIC  DC21041 (eth0)
 12:    11846      XT-PIC  PS/2 Mouse
 13:        1      XT-PIC  fpu
 14:   203376      XT-PIC  ide0
NMI:        0
```

Not surprisingly most of the numbers have gone up.

The Filesystem Hierarchy Standard (FHS)

A Linux system typically contains a very large number of files. For example a typical Red Hat installation may contain around 30,000 files occupying close to 400 Mbytes of disk space. Clearly it's imperative that these files be organized in some consistent, coherent manner. That's the motivation behind the filesystem hierarchy standard, FHS. The standard allows both users and software developers to "predict the location of installed files and directories"[3]. FHS is by no means specific to Linux. It applies to Unix-like operating systems in general.

The directory structure of a Linux filesystem always begins at the *root*, identified as "/". FHS specifies several directories and their contents directly subordinate to the root. This is illustrated in Figure 2-7. The FHS starts by characterizing files along two independent axes:

[3] *Filesystem Hierarchy Standard – Version 2.3* dated 1/29/04, edited by Rusty Russell, Daniel Quinlan, and Christopher Yeoh. Available from *www.pathname.com/fhs*.

the root directory

```
/ ─┐
   ├── bin     Essential command binaries
   ├── boot    Static files of the boot loader
   ├── dev     Device "inode" files
   ├── etc     Host-specific system configuration
   ├── home    Home directories for individual users (optional)
   ├── lib     Essential shared libraries and kernel modules
   ├── mnt     Mount point for temporarily mounting a filesystem
   ├── opt     Additional application software packages
   ├── root    Home directory for the root user (optional)
   ├── sbin    Essential system binaries
   ├── tmp     Temporary files
   ├── usr     Secondary hierarchy
   ├── var     Variable data
   │
```

Figure 2-7: Filesystem Hierarchy

- *Sharable vs. nonsharable.* A networked system may be able to mount certain directories through NFS such that multiple users can share executables. On the other hand, some information is unique to a specific computer and is thus not sharable.
- *Static vs. variable.* Many of the files in a Linux system are executables that don't change, they're *static*. But the files that users create or acquire, by downloading or e-mail for example, are *variable*. These two classes of files should be cleanly separated.

Here is a description of the directories defined by FHS:

- /bin – Contains binary executables of commands used both by users and the system administrator. FHS specifies what files /bin must contain. These include among other things the command shell and basic file utilities. /bin files are static and sharable.
- /boot – Contains everything required for the boot process except configuration files and the map installer. In addition to the kernel executable image, /boot contains data that is used before the kernel begins executing user-mode programs. /boot files are static and nonsharable.

- /etc – Contains host-specific configuration files and directories. With the exception of mtab, which contains dynamic information about filesystems, /etc files are static. FHS identifies three optional subdirectories of /etc:
  - /opt Configuration files for add-on application packages contained in /opt.
  - /sgml Configuration files for SGML and XML
  - /X11 Configuration files for X windows.
- In practice most Linux distributions have many more subdirectories of /etc representing optional startup and configuration requirements.
- /home (Optional) – Contains user home directories. Each user has a subdirectory under home with the same name as his/her user name. Although FHS calls this optional, in fact it is almost universal among Unix systems. The contents of subdirectories under /home is of course variable.
- /lib – Contains those shared library images needed to boot the system and run the commands in the root filesystem, i.e., the binaries in /bin and /sbin. In Linux systems /lib has a subdirectory, /modules, that contains kernel loadable modules.
- /mnt – Provides a convenient place to temporarily mount a filesystem. Typical subdirectories under /mnt are cdrom and floppy.
- /opt – Contains optional add-in software packages. Each package has its own subdirectory under /opt.
- /root – Home directory for the root user[4]. This is not a requirement of FHS but is normally accepted practice and highly recommended.
- /sbin – Contains binaries of utilities essential for system administration such as booting, recovering, restoring or repairing the system. These utilities are only used by the system administrator and normal users should not need /sbin in their path.
- /tmp – Temporary files.
- /usr – Secondary hierarchy, see below.
- /var – Variable data. Includes spool directories and files, administrative and logging data, and transient and temporary files. Basically, system-wide data that changes during the course of operation. There are a number of subdirectories under /var.

[4] Note the potential for confusion here. The directory hierarchy has both a root, "/", and a root directory, "/root", the home directory for the *root user*.

The /usr Hierarchy

/usr is a secondary hierarchy that contains user-oriented files. Figure 2-8 shows the subdirectories under /usr. Several of these subdirectories mirror functionality at the root. Perhaps the most interesting subdirectory of /usr is /src for source code. This is where the Linux source is generally installed. You may in fact have sources for several Linux kernels installed in /src under subdirectories with names of the form:

linux-<version number>-ext

You would then have a logical link named linux pointing to the kernel version you're currently working with.

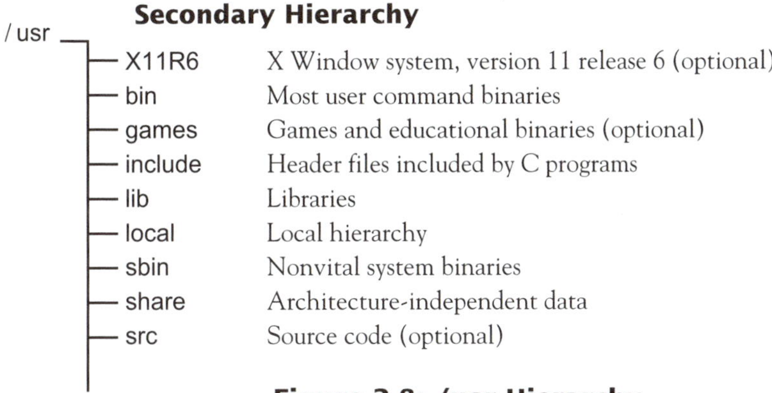

Figure 2-8: /usr Hierarchy

"Mounting" Filesystems

A major difference between Windows and Linux filesystems has to do with how file structured devices, hard disks, floppy drives, CDROMs, etc, are mapped into the system's directory or hierarchy structure. The Windows filesystem makes devices explicitly visible, identifying them with a letter-colon combination as in "C:". Linux on the other hand emphasizes a unified filesystem in which physical devices are effectively rendered invisible.

The mechanism that maps physical devices into the filesystem's directory structure is called *mounting*[5]. Removable media devices such as the CD-ROM drive are the most visible manifestation of this feature. Before you can read a CD-ROM, you must

[5] The term probably harks back to the days of reel-to-reel tape drives onto which a reel of tape had to be physically "mounted."

mount the drive onto an existing node in the directory structure using the mount command as in:

```
mount /mnt/cdrom
```

This command works because the file /etc/fstab has information about what device is normally mounted at /mnt/cdrom and what type of filesystem, in this case iso9660, the device contains.

Like file permissions, mount can sometimes be a nuisance if all you want to do is read a couple files off of a CD. But the real value of the mount paradigm is that it isn't limited to physical devices directly attached to the computer, nor does it only understand native Linux filesystems. As we'll see later, we can mount parts of filesystems on remote computers attached to a network to make their files accessible on the local machine. It's also possible to mount a device that contains a DOS FAT or VFAT filesystem. This is particularly useful if you build a "dual-boot" system that can boot up into either Windows or Linux. Files can be easily shared between the two systems.

System Configuration

The section above on the FHS mentioned the /etc directory. Here's one place where Unix systems really shine relative to Windows. Okay, there may be any number of ways that Unix outshines Windows. In any case, the /etc directory contains essentially all of the configuration information required by the kernel and applications in the form of plain text files. The syntax and semantics of these files isn't always immediately obvious, but at least you can read them. The format of many of the kernel's configuration files is documented in man pages (see section below on help).

By contrast, beginning with NT, Windows systems have made a point of hiding configuration information in a magical place called the *Registry*. Mortal users are often cautioned to stay away from the Registry because improper changes can render the system unbootable. In fact, the registry can only be changed by a special program, regedit, and it isn't even clear where the registry physically resides.

The Shell

One of the last things that happens as a Linux system boots up is to invoke the command interpreter program known as the *shell*. Its primary job is to parse commands you enter at the console and execute the corresponding program. But the shell is much more than just a simple command interpreter. It incorporates a powerful, expressive interpretive programming language of its own. Through a combination of the shell script language and existing utility programs it is quite possible to produce very sophisticated applications without ever writing a line of C code. In fact this is the general philosophy of Unix programming. Start with a set of simple utility programs and link them together through the shell scripting language.

The shell scripting language contains the usual set of programming constructs for looping, testing, functions, and so on. But perhaps the neatest trick in the shell bag is the concept of "pipes." This is a mechanism for streaming data from one program to another. The metaphor is perfect. One program dumps bytes into one end of the pipe while a second program takes the bytes out the other end.

Most Linux/Unix utility programs accept their input from a default device called *stdin*. Likewise, they write output to a device called *stdout*. Any error messages are written to "stderr." Normally, `stdin` is the keyboard while `stdout` and `stderr` are the display. But we can just as easily think of `stdin` and `stdout` as two ends of a pipe.

`stdin` and `stdout` can be *redirected* so that we can send output to, for example, a file or take input from a file. In a shell window, try typing just the command cat with no arguments. `cat`[6] is perhaps the simplest of all Unix utilities. All it does is copy `stdin` to `stdout`, line by line. When you enter the command with no arguments, it patiently waits at `stdin` for keyboard input. Enter a line of text and it will send that line to `stdout`, the display. Type Crtl + C to exit the program.

Now try this:

```
cat > textfile
```

Enter a line of text. This time you don't see the line repeated because the ">" operator has redirected stdout to the file `textfile`. Type a few lines, then Ctrl + C to exit.

The final step in this exercise is:

```
cat < textfile
```

Voila! The file you created with the previous command shows up on the screen because the "<" operator redirected stdin to `textfile`. cat actually implements a shortcut so that if you enter a filename as a command line argument, without the < operator, it takes that as an input file. That is:

```
cat textfile
```

[6] Short for con**cat**enate. Don't you just love Unix command names?

The real power of pipes though is the " | " operator, which takes stdout of one program and feeds it to stdin of another program. When I did the above exercises, I created a textfile containing:

```
this is a file
of text from the keyboard
```

Now if I execute:

```
cat textfile | grep this
```

I get:

```
this is a file
```

grep, as you may have guessed, is a filter. It attempts to match its command line arguments against the input stream stdin. Lines containing the argument text are passed to stdout. Other lines are simply ignored. What happened here is that the output of cat became the input to grep.

In typical Unix fashion, grep stands for Get Regular Expression "something." I forget what the "p" stands for. Regular expressions are in fact a powerful mechanism, a scripting language if you will, for searching text. grep embodies the syntax of regular expressions.

Shell scripts and makefiles make extensive use of redirection and piping.

Two other shell features are worth a brief mention because they can save a lot of typing. The shell maintains a command history that you can access with the up arrow key. This allows you to repeat any previous command, or edit it slightly to create a new, similar command. The other cool feature is autocompletion, which attempts to complete filename arguments for you. Say I wanted to execute

```
cat verylongtextfilename
```

I could try entering

```
cat verylong<tab>
```

Provided the remainder of the filename is unique in the current directory, the shell will automatically complete it, saving me the trouble of typing the whole filename. The shell beeps if it can't find a unique match. Then you just have to type a little more until the remainder is unique.

There are in fact several shell programs in common use. They all serve the same basic purpose yet differ in details of syntax and features. The most popular are:

- *Bourne Again Shell* – bash. bash is a "reincarnation" of the Bourne shell, sh, originally written by Stephen Bourne and distributed with Unix version 7. It's the default on most Linux distributions and you should probably use it unless you have good reason, or strong feelings, for using a different shell.
- *Korn Shell* – ksh. Developed by David Korn at AT&T Bell Laboratories in the early 1980s. It features more advanced programming facilities than bash but nevertheless maintains backward compatibility.
- *Tenex C Shell* – tcsh. A successor to the C shell, csh that was itself a predecessor to the Bourne shell. Tenex was an operating system that inspired some of the features of tcsh.
- *Z Shell* – zsh. Described as an extended Bourne shell with a large number of improvements, including some of the most useful features of bash, ksh, and tcsh.

The subject of shell programming is worthy of a book in itself and there are many. When I searched Amazon.com for "shell programming," I got 189 hits.

Getting Help

The official documentation for a Unix/Linux system is a set of files called *man pages*, man being short for manual. man pages are accessed with the shell command man as in:

```
man cp
```

to get help on the shell copy command. Try it. Type man man at the shell prompt to learn more about the man command. man presents the page a screen at a time with a ":" prompt on the bottom line. To move to the next screen, type <space>. To exit man before reaching the end of the page, type "q". Interestingly enough, you won't find that information in the man man page. The writing style in man pages is rather terse, they're reference in nature, not tutorial. The information is typically limited to: what the command does, what its arguments are, and what options it takes.

To make it easier to find a specific topic the man pages are organized into sections as follows:

Section 1:	User Commands entered at the shell prompt.
Section 2:	The kernel API.
Section 3:	C library functions.
Section 4:	Devices. Information on specific peripheral devices
Section 5:	File formats. Describes the syntax and semantics for many of the files in /etc.
Section 6:	Games.
Section 7:	Miscellaneous.
Section 8:	System Administration. Shell commands primarily used by the system administrator.

Resources

Sobell, Mark G., *A Practical Guide to Linux*. This book has been my bible and constant companion since I started climbing that steep Linux learning curve. It's an excellent beginner's guide to Linux and Unix-like systems in general, although having been published in 1997 it is getting a bit dated and hard to find. It has been superceded by…

Sobell, Mark G., *A Practical Guide to Linux Commands, Editors, and Shell Programming*

tldp.org – The Linux Documentation Project. As the name implies, this is the source for documentation on Linux. You'll find how-to's, in-depth guides, FAQs, man pages, even an online magazine, the *Linux Gazette*. The material is generally well-written and useful.

CHAPTER 3

The Host Development Environment

"If Bill Gates had a nickel for every time Windows crashed…Oh wait, he does"

—Spotted on Slashdot.org

In many cases, the target computer on which an embedded application runs is severely limited in terms of its computing resources. It probably doesn't have a full-sized display or keyboard. It may have at most a few megabytes of mass storage in the form of a flash file system, hardly enough to contain a compiler much less a decent integrated development environment (IDE). Thus embedded development usually requires at least two computers—the target on which the embedded program will run and a development workstation on which the embedded program is written and compiled. Before we begin working with an embedded Linux environment, we'll have to set up an appropriate development workstation.

Any modern PC will work just fine as a development host. Minimum requirements are: a Pentium class processor, 128 Mbytes of RAM for graphical operation, and 2 Gbytes of disk for a "workstation" class Linux installation. Of course, more RAM is always better.

I do most of my Linux work on a 300 MHz AMD K6 with 224 MB of RAM that I call my "lab computer" (see Figure 3-1). It's a dual-boot system since I also work in the (shudder!) Microsoft world. I have 2 Gbytes of disk dedicated to Win 95[1] and 6 Gbytes for Linux, of which about a third is currently used. I currently run Red Hat version 9.0.

1 "Win 95!," I hear you gasp. Emphatically yes. I still occasionally knock out a 16-bit DOS program to do simple hardware testing. The DOS window in Win 95 lets me directly read and write peripheral registers. That becomes increasingly difficult, if not impossible, in later versions of Windows.

You will need at least one, and preferably two, asynchronous serial ports. You will also need a network interface. OK, you might be able to live without it, but it does make life a lot easier. We'll use a combination of serial and network ports to communicate with the target as well as debug target code.

Figure 3-1: My Linux Development Machine

Linux Distributions

There are any number of complete Linux "distributions" floating around. A distribution is a collection of "packages" implementing the various functions required in a full operating system—kernel, shell, utilities, graphical environment, development tools, office tools, games, and so on. A typical distribution includes a couple hundred packages organized into a dozen or so categories. Some people rate Linux distributions by the number of packages they contain ("My Linux distribution has more packages than your Linux distribution").

In most distributions, packages are in "RPM format" where RPM means Red Hat package manager. RPM is an Open Source packaging system developed by Red Hat that has become something of a de facto standard for distributing software. It's a powerful utility that keeps track of where software packages should be installed, the versions of packages you have installed, and the dependencies among packages.

RPM maintains a database of installed packages that tracks what files belong to what packages and their dependencies. One consequence of this is that you can verify the integrity of a package. Are all the files there? If not, reinstall it. Another consequence is that software installed with RPM should only be removed with RPM. That is, don't just go in and delete the files. In this sense, RPM is something like the Add/Remove Programs item in the Windows Control Panel.

While you're free to install your favorite Linux distribution, be it Red Hat, Debian, Suse, Mandrake, whatever, I recommend either Red Hat 8 or Red Hat 9 for a couple of reasons. The BlueCat embedded Linux package that we'll be working with in the next two chapters has been tested and verified with Red Hat 8 and Red Hat 9 and those are the only ones that LynuxWorks will support.

The other reason for choosing a fairly recent distribution, such as Red Hat 9, has to do with the 2.6 series kernel. Until recently, my favorite distribution was Red Hat 7.2, based on the 2.4.10 kernel. It's quite stable, fairly capable, and appears to play nice with BlueCat version 5. Then I tried building and running a 2.6 kernel. Turns out there are so many incompatibilities between a 2.4 environment and a 2.6 environment in terms of tool chains, scripts, and so on that it just isn't worth the hassle. You want a distribution that's based on a 2.6 series kernel.

Modern Linux distributions sport reasonably clean, self-explanatory installation procedures, so I won't go into great detail on that here. But a few cautionary observations are in order. You will typically get a choice of "server," "workstation," or "custom" installation. While a workstation class installation should be adequate, I generally do a custom installation to be sure I get exactly the tools I need without cluttering up my disk with a bunch of stuff I don't need. You'll need the development tools of course. You don't need the kernel source because we'll be installing that from the book CD in the next chapter.

If you're upgrading from a previous Linux distribution, one of the options is to "upgrade" rather than do a complete new installation. The upgrade preserves the existing file structure and simply replaces packages as necessary. But my experience in upgrading from Red Hat 7.2 to 9.0 was that it seemed to leave the system in an unstable state where things didn't work quite right. I finally did a clean install.

Dual-Booting

If you happen to have a computer that you can dedicate as a Linux workstation, that's great. It's probably the simplest environment to run in. In many cases though, you'll probably want to install Linux on a machine that already runs some variant of Windows. This is known as *dual-booting*.

The first requirement for creating a dual-boot environment on top of Windows is to defragment the disk so that all the free space is at the "end." Then you need to create a new "partition" on the disk to hold Linux. There are many tools for doing this. Linux distributions typically include an Open Source tool called FIPS although my experience has been that it doesn't get along well with later versions of Windows. Probably the best disk partitioning tool is Partition Magic, which you can buy off the Internet for less than $20. Google "partition magic" to find a vendor.

With a new partition in place, you can proceed to Linux installation. When it gets to the point of setting up a bootloader, you should see the original Windows system and the new Linux system listed as boot options. Next time you boot the machine you'll see a screen with both systems listed and one selected as a default. If you do nothing, the default system will boot, typically after 5 seconds.

An alternative to the above scenario is a commercial package called VMware that creates one or more "virtual machines" on an x86 system. In effect, the virtual machine is a very large file in the native system's filesystem. VMware's virtualization layer maps the physical hardware resources to the virtual machine's resources, so each virtual machine has its own CPU, memory, disks, and I/O devices, and is the full equivalent of the native system.[2]

The neat thing about the VMware approach is that both the native system and virtual system are running at the same time. You can switch back and forth between

---

[2] Although in my experimentation with VMware, I found that a Linux device driver could not directly read and write the parallel port. So it appears that I/O device access is at a more abstract level.

them and even drag and drop files between systems. The downside is the retail price for VMware Workstation is $189, although you can find it for a little less on the net. A 30-day free evaluation version is available. See the Resources section later in the Chapter.

Cross-Development Tools—The GNU Tool Chain

Not only is the target computer limited in resources, it may be a totally different processor architecture from your (probably) x86-based development workstation. Thus we need a cross-development tool chain that runs on the PC but may have to generate code for a different processor. We also need a tool that will help us debug code running on that (possibly different) target.

GCC

By far the most widely used compiler in the Linux world is GCC, the gnu compiler collection. It was originally called the gnu C compiler but the name was changed to reflect its support for more than just the C language. GCC has language front ends for C, C++, Objective C, Fortran, Java, and Ada as well as run-time libraries for these languages.

GCC also supports a wide range of target architectures in addition to the x86. Supported targets include:

- Alpha
- ARM
- M68000
- MIPS
- PowerPC
- SPARC

and probably others that I'm not aware of.

GCC can run in a number of operating environments including Linux and other variants of Unix. There are even versions that run under DOS and Windows.

GCC can be built to run on one architecture (a PC for example) while generating code for a different architecture (an ARM perhaps). This is the essence of cross development.

GCC is both compiler and linker. For a program consisting of a single C source file, the command to build an executable can be as simple as:

```
gcc myprogram.c
```

By default this creates an executable called a.out. The –o option lets you do the obvious thing and rename the executable thus:

```
gcc –o myprogram myprogram.c
```

If your project consists of multiple source files, the –c option tells GCC to compile only.

```
gcc –c myprogram.c
```

generates the relocatable object file **myprogram.o**. Later this can be linked with other .o files with another GCC command to produce an executable.

Generally though, you won't be invoking GCC directly. You'll use a Makefile instead. See below.

GDB

GDB stands for the "gnu debugger." This is a powerful source-level debugging package that lets you see what's going on inside your program. You can step through the code, set breakpoints, examine and change variables and so on. Like most Linux tools, GDB itself is command line driven, making it rather tedious to use. There are several graphical front ends for GDB that translates graphical user interface (GUI) commands into GDB's text commands. We'll look at one of those, DDD, in Chapter 6.

GDB can be set up to run on a host workstation while debugging code on a separate target machine. The two machines can be connected via serial or network ports, or you can use an in-circuit emulator (ICE). Many ICE vendors have GDB back ends for their products.

There are also specialized debug ports built into some processor chips. These go by the names JTAG (Joint Test Action Group) and BDM (Background Debug Mode). These ports provide access to debug features such as hardware breakpoint registers. GDB back ends exist for these ports as well.

Make

Real world projects consist of anywhere from dozens to thousands of source files. These files have dependencies such that if you change one file, a header for example, another file, a C source, needs to be recompiled and ultimately the whole project needs to be rebuilt. Keeping track of these dependencies and determining what files need to be rebuilt at any given time is the job of a powerful tool called the Make utility.

In response to the shell command make, the Make utility takes its instructions from a file named **Makefile** or **makefile** in the current directory. This file consists of *dependency* lines that specify how a *target file* depends on one or more *prerequisite files*. If any of the prerequisite files are more recent than the target, make updates the target using one or more *construction commands* that follow the dependency line. The basic syntax is:

```
target: prerequisite-list
<tab> construction commands
```

But don't let this simple example fool you. There is a great deal of complexity and subtlety in the **make** syntax, which looks something like the shell scripting language. To get a feel for just how complex it can get, take a look at the **Makefile** in **/usr/src/linux** after you've untarred the kernel sources in the next chapter.

Often the best way to create a Makefile for a new project is to start with the one for an existing, similar project.

Resources

www.linuxiso.org – This site has downloadable ISO image files for virtually all of the popular Linux distributions on several architectures at various revision levels. An ISO file is an exact image of a CD or DVD. Note that burning an ISO image to a CD is not the same as copying a file. Many popular CD burning programs have an option for burning ISO images.

Bear in mind that a complete Linux distribution occupies two to four CDs. Unless you have a broadband Internet connection, don't even think about trying to download one.

www.vmware.com – Information about the VMware virtual machine. Among other things, you can download a free 30-day evaluation version.

Mecklenburg, Robert, *Managing Project with GNU Make 3rd Edition*, O'Reilly, 2005. This is the bible when you're ready to fully understand and exploit the power of make.

CHAPTER 4

Configuring and Building the Kernel

"Hackito ergo sum"
—*Anonymous*

One of the neatest things about Linux is that you have the source code. You're free to do whatever you want with it. Most of us have no intention, or need, to dive in and directly hack the kernel sources. But access to the source code does mean that the kernel is highly configurable. That is, you can build a kernel that precisely matches the requirements, or limitations, of your target system.

Now again, if your role is writing applications for Linux, or if you're a Linux system administrator, you may never have to touch the kernel. But as an embedded-systems developer, you will most certainly have to build a new kernel, probably several times, either for the workstation or the target environment. Fortunately, the process of configuring and building a kernel is fairly straightforward. My experience has been that building a new kernel is a great confidence building exercise, especially if you're new to Linux.

When building a new kernel, it's generally a good idea to start with clean, "virgin," source downloaded from *www.kernel.org*, particularly if you need to patch the kernel to support some added functionality such as real-time application interface (RTAI). Patch files are generally based on clean kernel sources and may not execute correctly on sources that have been otherwise modified.

The remainder of this chapter details and explains the various steps required to build an executable kernel image.

Getting Started

The book CD includes a Bzip file of the Linux kernel version 2.6.7. The file is `linux-2.6.7.tar.bz2` in the directory `/rtai`. Copy this file to `/usr/src` and unzip it. First though, check to be sure /usr and /src have read and write permission for "other" enabled so that you can work with the kernel as a normal user.

If you're like me, you'll just open the file with the KDE archiver, which will automatically recognize the file format and choose the appropriate decompression utility. If you insist on using a shell command, cd to /usr/src, if you're not there already, and execute

```
bunzip2 linux-2.6.7.tar.bz2
tar –xf linux-2.6.7.tar
```

Astute readers may have noticed that the latest stable kernel version at kernel.org is "somewhat higher" than 2.6.7. At the time this is being written the latest stable release is 2.6.13.1 dated 9/10/2005. 2.6.7 is dated 6/15/2004. Why pick such an "old" version? I simply choose to stay away from the "bleeding edge" and stick with something that is *really* stable. Heck, some vendors are still shipping version 2.4 kernels. I'm not quite that conservative because there are significant new features in the 2.6 series and this book would be seriously deficient if it didn't address them.

The specific reason for choosing 2.6.7 is that later in the book we'll be looking at the RTAI real-time extensions to Linux. The most stable version of RTAI, at the time this is being written, is 3.1, which happens to have patches for the 2.6.7 kernel.

By all means, feel free to play around with any versions of the kernel that strike your fancy. That's what they're there for. But when it comes time to ship a product, stability of the software is critical. Stick with a version that's been around for a while.

Where is the Source Code?

Following the steps in the previous section, you'll find a new directory in /usr/src, linux-2.6.7. It's not at all unusual to have several versions of the Linux kernel and corresponding source code on your system. How do you cope with these multiple versions? There is a simple naming strategy that makes this complexity manageable.

Kernel Version Numbering

Generally, Linux sources are installed as subdirectories of /usr/src. The subdirectories usually get names of the form

```
linux-<version_number>-<additional_features>.
```

This gives us an excuse to digress into how kernel versions are numbered. <version_number> identifies the base version of the kernel as obtained from kernel.org and looks something like this: 2.6.7. The first number is the "version," in this case 2. This number increments only when truly major architectural changes are made to the kernel and its application program interface (API). We've been at 2 for something like eight years now[1].

The second number, 6, is called the "patch level" and identifies sub-versions where the kernel API may differ but the differences aren't enough to justify a full version change. Up until the release of the 2.6 kernel, there was an interesting policy about patch levels. Even numbers represented stable, production kernels and odd numbers identified the kernel currently under development.

The final number, 7 in this example, is called the "sublevel." Basically it represents bug fixes and enhancements that don't change the kernel API. Applications built to a specific patch level should, in principle, run on any sublevel.

<additional_features> is a way of identifying patches made to a stock kernel to support additional functionality. We've installed a stock 2.6.7 kernel and so the <additional_features> field is null. Later on we'll patch this kernel to support RTAI and we'll create a new directory tree with the appropriate <additional_features> field.

The version numbering strategy changed beginning with version 2.6. There is no 2.7 development version. Instead there are now three version 2.6 kernel "trees." The first is the mainline or stable kernel, known as *2.6.x*. The second is the developmental tree, called 2.6-mm, where new technologies are tested before being merged into the mainline tree. Finally, there's a tree for bug fixes only, known as *2.6.x.y*. The .y code is folded into the next increment of 2.6.x and .y starts over again.

The primary objective of this new strategy is to make kernel updates more "gradual." The jump from 2.4 to 2.6 was somewhat painful and vendors in particular appreciate having changes appear in smaller increments.

Whatever its subdirectory name, the kernel you're currently working with is identified by a symbolic link in /usr/src called linux. Create the symbolic link with this shell command:

```
ln –s linux-2.6.7 linux
```

[1] At the time version 2.6 was released in early 2004, there was talk of bumping it to 3.0, but the decision was made to stick with 2.6. One of the rationales for staying the course was that a jump in major number implies binary imcompatibility, but 2.6 is binary compatible 2.4.

The Kernel Source Tree

Needless to say, the kernel encompasses a very large number of files—C sources, headers, makefiles, scripts, etc. So not surprisingly, there's a standard directory tree to organize these files in a manageable fashion. Figure 4-1 shows the kernel source tree starting at /usr/src/linux. The directories are as follows:

Documentation – Pretty much self-explanatory. This is a collection of text files describing various aspects of the kernel, problems, "gotchas," and so on. There are several subdirectories under Documentation for topics that require more extensive explanations.

arch – All architecture-dependent code is contained in subdirectories of arch. Each architecture has a directory under arch with its own subdirectory structure. The executable kernel image will end up in arch/<architecture>/boot. An environment variable in the Makefile, $ARCH, points to the appropriate target architecture directory.

crypto – Code dealing with the cryptographic aspects of system security.

drivers – Device driver code. Under drivers is a set of subdirectories for various devices and classes of device.

fs – Filesystems. Under fs is a set of directories for each type of filesystem that Linux supports.

include – Header files. Among the subdirectories of include are a set of the form "asm-<arch>" where <arch> is the same name as the subdirectory of arch that represents a specific architecture. These directories hold header files containing in-line assembly code, which of course is architecture-dependent. A link named asm points to the subdirectory of the target you're building for.

init – Contains two files; main.c and version.c.

ipc – Code to support Unix System 5 Interprocess Communication mechanisms such as semaphores, message passing and shared memory.

kernel – This is the heart of the matter. Most of the basic architecture-independent kernel code that doesn't fit in any other category is here.

lib – Several utility functions that are collected into a library.

mm – Memory management functions.

net – Network support. Subdirectories under net contain code supporting various networking protocols.

scripts – Text files and shell scripts that support the configuration and build process.

security – Offers alternative security models for the kernel.

sound – Support for sound cards and the Advanced Linux Sound Architecture.

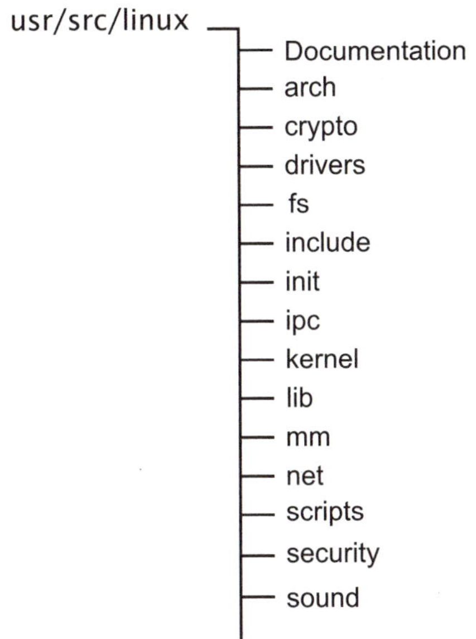

Figure 4-1: Linux Source Tree

Configuring the Kernel—make config, menuconfig, xconfig

usr/src/linux contains a standard Makefile, Makefile, with a very large number of make targets. The process of building a kernel begins by invoking one of the three make targets that carry out the configuration process. make config starts a text-based script that sequentially steps you through each configuration option. For each option you have either three or four choices. The three choices are: "y" (yes), "n" (no), and "?" (ask for help). The default choice is shown in upper case.

Some options have a fourth choice, "m", which means build this feature as a loadable kernel module rather than build it into the kernel image. Kernel modules are a way of dynamically extending the kernel and are particularly useful for things like device drivers. Chapter 7 goes into modules in some detail.

Figure 4-2 shows an excerpt from the make config dialog.

Most options include help text that is generally "helpful" (See Figure 4-3).

The problem with make config is that it's just downright tedious. Typically you'll only be changing a very few options and leaving the rest in their default state. But make config forces you to step through each and every one. Personally I've never used make config and I wouldn't recommend it.

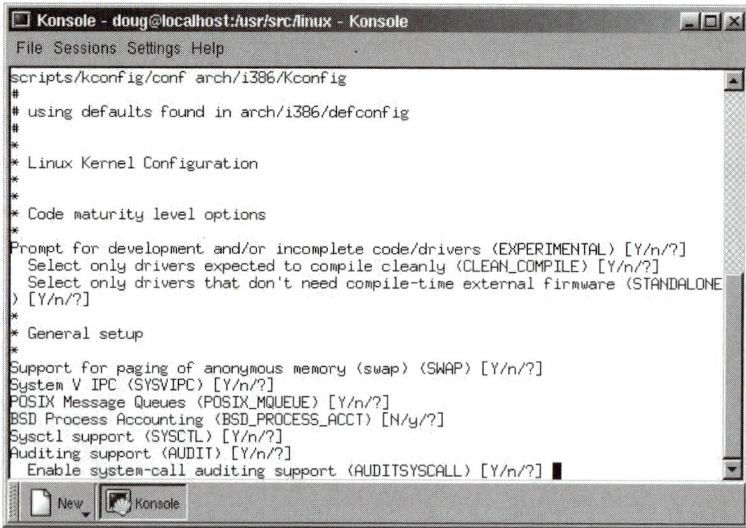

Figure 4-2: make config Dialog

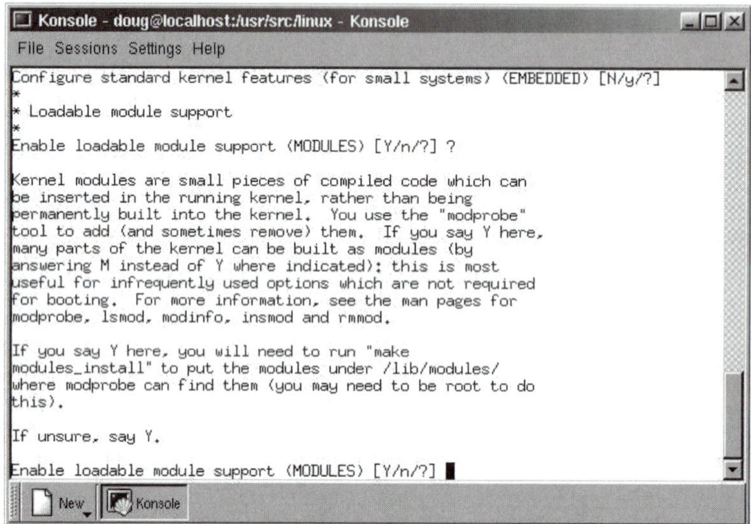

Figure 4-3: make config Help Text

make menuconfig brings up the pseudo-graphical screen shown in Figure 4-4. Here the configuration options are grouped into categories and you only need to visit the categories of options you need to change. The interface is well explained and reasonably intuitive. The same help text is available as with make config. When you exit the main menu, you are given the option of saving the new configuration.

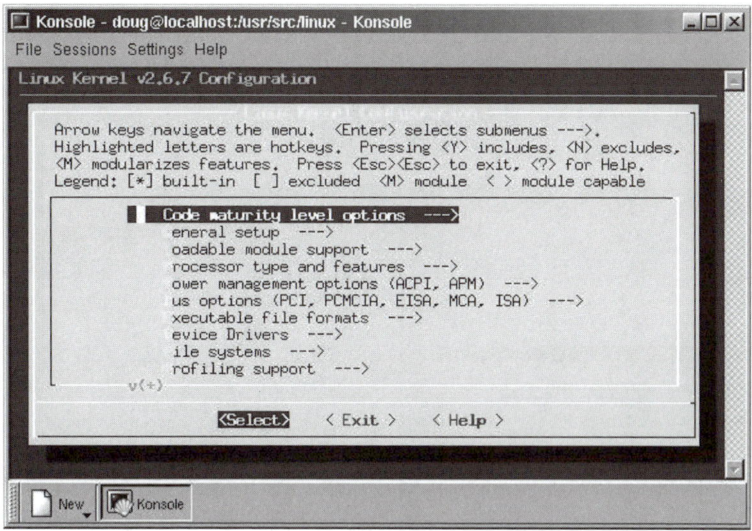

Figure 4-4: make menuconfig Main Menu

But my choice for overall ease of use is make xconfig. This brings up an X Windows-based menu as shown in Figure 4-5. Now you can see all the option categories at once and navigate with the mouse. Of course you must be running X Windows to use this option.

make xconfig has improved substantially from kernel version 2.4 to 2.6, but if you've never built a kernel before you can't truly appreciate the improvements. Principal among them is the help window in the lower right hand corner that always displays help text for whatever configuration option is selected.

> **Caution:** *There are a couple of potential "gotchas" in xconfig for the 2.6 series. The compiler version recommended for building the kernel, 2.95.3, fails to build the xconfig programs. The 3.2.2 compiler that ships with Red Hat 9.0 does work. Secondly, xconfig now requires the Qt library, which may or may not be installed by default. I had to install K desktop environment (KDE) development tools to get it.*

In its default display mode, **xconfig** displays a tree of configuration categories on the left. Selecting one of these brings up the set of options for that category in the upper right window. Figure 4-5 is an excerpt from the Processor Type and Features menu. Most of the options are presented as check boxes. Clicking on the box alternates between checked (selected) and unchecked (not selected). Options that may be built as kernel modules have a third value, a dot to indicate that the feature will be built as a module, for example, /dev/cpu/microcode.

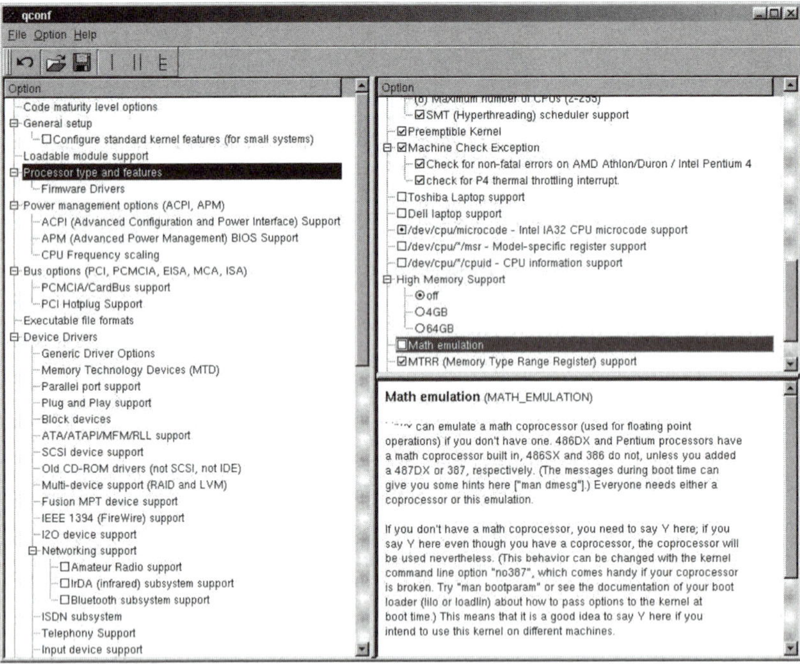

Figure 4-5: make xconfig

> **Try it out**
>
> The best way to get familiar with kernel configuration options is to fire up xconfig and see what's there. So...
>
> ```
> cd /usr/src/linux
> make xconfig
> ```
>
> After a bit of program and file building, the menu of Figure 4-5 will appear. Just browse through the submenus and their various sub-submenus to get a feel for the flexibility and richness of features in the Linux kernel. Read the help descriptions.
>
> Under File systems → Network File Systems, ensure both NFS File system support and NFS server support are enabled. Other than that, you'll want to verify that drivers for your particular hardware are enabled.
>
> By default, a large number of driver options are enabled as modules. Chances are your system supports only a small fraction of this hardware. Disabling the modules you don't need will speed the kernel build process.
>
> For the time being leave the configuration menu up while you read the next section.

Some options like "High Memory Support" have a set of allowable values other than yes or no and are represented by radio buttons. Some options take numeric values. Options that aren't available, because some other option was not selected, simply don't show up. As an example, under `Device Drivers` in the left hand pane select `Memory Technology Devices (MTD)`. The right hand window should show an unchecked box for Memory Technology Device (MTD) Support and nothing else. If you check this box, you'll see a number of additional options show up in both the right and left-hand windows.

`make xconfig` gives you additional flexibility in saving configurations. Instead of always saving it to the standard `.config` file, you can save it to a named file of your choice using the File->Save As option and later load that file using File->Load for further modification or to make it the new configuration.

Behind the Scenes—What's Really Happening

The information in this section is not essential to the process of building a kernel and you're free to ignore it for now. When you reach the point of developing device drivers or other kernel enhancements (or perhaps hacking the kernel itself), you will need to modify the files that control the configuration process.

```
menu "Processor type and features"

choice
   prompt "Subarchitecture Type"
   default X86_PC

config X86_PC
   bool "PC-compatible"
   help
      Choose this option if your computer is a standard PC or
      compatible.

config X86_ELAN
   bool "AMD Elan"
   help
      Select this for an AMD Elan processor.

      Do not use this option for K6/Athlon/Opteron processors!

      If unsure, choose "PC-compatible" instead.

config X86_VOYAGER
   bool "Voyager (NCR)"
   help
      Voyager is a MCA based 32 way capable SMP architecture
      proprietary to NCR Corp. Machine classes 345x/35xx/4100/
      51xx are voyager based.
```

Listing 4-1: Excerpt of Kconfig

All the information in the configuration menus is provided by a set of text files named Kconfig. These are script files written in Config Language, which looks suspiciously like a shell scripting language but isn't exactly. The main Kconfig file is located in linux/arch/$(ARCH) where ARCH is a variable in Makefile identifying the base architecture. Listing 4-1 is an excerpt from the processor type and features submenu of the i386 architecture.

Go to linux/arch/i386 and open Kconfig. Compare the structure of the file with the configuration menu and the pattern should become fairly clear. Each configuration option is identified by the keyword config followed by the option's symbolic name, for example config X86_PC. Each of these gets turned into a Makefile variable, or macro, such as CONFIG_X86_PC. The makefiles then use these variables to determine which components to include in the kernel and to pass #define symbols to the source code.

The option type is indicated by one of the following keywords:

- *bool* – The option has two values, "y" or "n".
- *tristate* – The option has three values, "y", "n", and "m".
- *choice* – The option has one of the listed values.
- *int* – The option takes an integer value.

The type keyword usually includes a prompt, which is the text displayed in the configuration menu. Alternatively the prompt keyword specifies the prompt. There are also "hidden" options that have no displayable prompt. At the top of Kconfig is a set of options that don't show up in the menu because no prompt is specified. In the configuration menu, click on Options->Show All. Now all options are displayed, including those with no prompt and those dependent on unselected options.

Other keywords within a configuration entry include help to specify the help text and depends, which means that this option is only valid, or visible, if some other option that it depends on is selected. Config Language is actually much more extensive than this simple example would suggest. For more detail, look at linux/Documentation/kbuild/kconfig-language.txt.

Just above the menu "Processor type and features" line shown in Listing 4-1 is the line:

```
source init/Kconfig
```

The `source` keyword is how the configuration menu is extended to incorporate additional, modular features. You'll find `source` lines scattered throughout the main Kconfig file.

The end result of the configuration process, whichever one you choose to use, is a file called .config containing all of the Makefile variables. Listing 4-2 is an excerpt. The options that are not selected remain in the file but are "commented out." The selected options get the appropriate value, "y", "m", a number, or an element from a list. .config is included in the Makefile.

```
#
# Automatically generated make config: don't edit
#
CONFIG_X86=y
CONFIG_MMU=y
CONFIG_UID16=y
CONFIG_GENERIC_ISA_DMA=y

#
# Code maturity level options
#
CONFIG_EXPERIMENTAL=y
CONFIG_CLEAN_COMPILE=y
CONFIG_STANDALONE=y

#
# General setup
#
CONFIG_SWAP=y
CONFIG_SYSVIPC=y
# CONFIG_POSIX_MQUEUE is not set
# CONFIG_BSD_PROCESS_ACCT is not set
```

Listing 4-2: Excerpt of .config

Building the Kernel

I highly recommend building the 2.6 kernel under a Linux distribution based on a 2.6 kernel. While it is theoretically possible to build a 2.6 kernel under an older distribution, many utilities need to be updated and any number of things can go wrong. But if you insist on trying, Appendix D summarizes all of the relevant version information and where to find updates if needed.

Despite the kernel developer's seemingly outdated recommendation to use gcc version 2.95.3[2], I use the 3.2.2 compiler that ships with Red Hat™ 9.0 because I insist on using `xconfig` to configure the kernel. To all appearances 3.2.2 builds a working 2.6.7 kernel. A binary distribution of gcc 2.95.3 is on the book CD as `/tools/gcc-2.95.3bin.tar.gz`. If needed, untar this file in `/usr/local` and be sure `/usr/local/bin` is at the front of your PATH.

The actual process of building the kernel has been simplified somewhat since the 2.4 series. The first two steps below can be executed as a normal user. The build process comprises the following steps:

> *make clean*. Deletes all intermediate files created by a previous build. This insures that *everything* gets built with the current configuration options. You'll find that virtually all Linux makefiles have a `clean` target. Strictly speaking, if you're building for the first time, there are no intermediate files to clean up so you don't have to run `make clean`.
>
> *make*. This is the heart of the matter. This builds the executable kernel image and all kernel modules. Not surprisingly, this takes a while. The resulting compressed kernel image is `arch/$(ARCH)/boot/bzImage`.

The following steps require root user privileges:

> *make modules_install*. Copies the modules to `/lib/modules/<kernel_version>` where `<kernel_version>` is the string identifying the specific kernel you are building.
>
> *cp* `arch/$(ARCH)/boot/bzImage /boot/vmlinuz-<kernel_version>`. This copies the kernel image to the `/boot` directory. As written here, it assumes you're currently in the directory `/usr/src/linux`. Actually, you can name the target file in `/boot` anything you want, but this is the convention and just makes things easier.

---

[2] In part because that's what Linus uses.

cp System.map /boot/System.map-<kernel_version>. Copies the system map file to /boot.

Note incidentally that the build process is recursive. Every subdirectory in the kernel source tree has its own Makefile dealing with the source files in that directory. The top level Makefile recursively invokes all of the sub makefiles.

There's one last step. Most Linux kernels are set up to use initrd, which is short for initial ramdisk. An initrd is a very small Linux filesystem loaded into RAM by the boot loader and mounted as the kernel boots, before the main root filesystem is mounted. The usual reason for using initrd is that some kernel modules need to be loaded before mounting the root partition. Usually these modules are required to support the filesystem used by the root partition, ext3 for example, or perhaps the controller that the hard drive is attached to, such as SCSI or RAID.

Without going into a lot of detail, the upshot is that we need to create an initrd image that gets mounted as the initrd. Fortunately, this is fairly simple, at least under Red Hat distributions. Just execute the command

```
/sbin/mkinitrd /boot/initrd-<kernel_version>.img <kernel_version>
```

The first argument, the image file name is arbitrary. What's shown is the naming convention. mkinitrd is a script that knows what sorts of modules need to be included in the initrd. Once you create the initrd image it's valid for subsequent kernel builds unless you change the root filesystem or the device it boots from.

Booting the New Kernel

Most Linux installations incorporate a boot loader, either LInux LOader (LILO) or GRand Unified Bootloader (GRUB) to select the specific kernel or alternate operating system to boot. There's a very good reason for having the ability to boot multiple kernel images. Suppose you build a new kernel and it fails to boot properly. You can always go back and boot a known working image and then try to figure out what went wrong in your new one.

We now need to add our new kernel to the list of kernels recognized by the bootloader. Information about the kernels that LILO can boot is kept in /etc/lilo.conf. Listing 4-3 is an example lilo.conf file. This example shows the standard Red Hat 9.0 Linux kernel image, 2.4.20-8, along with an "other" operating system named DOS. If no image is specified at the LILO prompt, the default is DOS after 5 seconds.

```
prompt
timeout=50
default=DOS
boot=/dev/hda
map=/boot/map
install=/boot/boot.b
message=/boot/message
linear

image=/boot/vmlinuz-2.4.20-8
    label=linux
    initrd=/boot/initrd-2.4.20-8.img
    read-only
    root=/dev/hda3

other=/dev/hda1
    optional
    label=DOS
```

Listing 4-3: lilo.conf

The easiest way to add a new kernel to lilo.conf is to just copy and paste the section for an existing kernel, the five lines beginning with "image =". Then change the image file name, the initrd name, and the label to match the kernel you've just built. After saving the file, you must run the command lilo to actually install the boot loader. If you're not logged in as the root user you may have to type /sbin/lilo because /sbin is typically not in the path of a normal user.

The process with GRUB is much the same. GRUB's boot information is kept in /boot/grub/grub.conf, an example of which is shown in Listing 4-4. In this case, copy and paste the section that begins with "title" up to but not including the next line that begins with "title." Unlike LILO, GRUB does not need to be run to install the boot loader. It gets its information directly from grub.conf at boot time.

```
# grub.conf generated by anaconda
#
# Note that you do not have to rerun grub after making changes to
this file
# NOTICE:  You have a /boot partition.  This means that
#          all kernel and initrd paths are relative to /boot/, eg.
#          root (hd0,1)
#          kernel /vmlinuz-version ro root=/dev/hda3
#          initrd /initrd-version.img
#boot=/dev/hda
default=1
timeout=10
splashimage=(hd0,1)/grub/splash.xpm.gz
title Red Hat Linux (2.4.20-8)
    root (hd0,1)
    kernel /vmlinuz-2.4.20-8 ro root=/dev/hda3
    initrd /initrd-2.4.20-8.img
title Windows 95
    rootnoverify (hd0,0)
    chainloader +1
```

Listing 4-4: grub.conf

What Can Go Wrong?

Plenty! I said at the beginning of this chapter that configuring and building a kernel is "fairly straightforward." That was true with the 2.4 series. And it's also true if you're building a 2.6 kernel from within a distribution based on a 2.6 kernel. But right now, late 2005, we're still in a transition phase with many people using 2.4-based distributions because certain packages, like BlueCat™ Linux depend on that.

The bottom line is you are likely to encounter some difficulties in building a 2.6 kernel. It took me about 3 weeks of futzing around to finally get a 2.6 kernel to boot correctly. I've flagged some of the problems I encountered, but you may find others.

If all else fails, you can fall back on a 2.4 kernel for your workstation. Use the one that came with your distribution or try my personal favorite, 2.4.18.

Resources

For more details on the process of configuring and building a kernel, look at the files in /usr/src/linux/Documentation/kbuild. Additionally, the following HOWTOs at *www.tldp.org* may be of interest:

Config-HOWTO – This HOWTO is primarily concerned with how you configure your system once it's built.

Kernel-HOWTO – Provides additional information on the topics covered in this chapter.

CHAPTER 5

BlueCat Linux

"To err is human.
To really foul things up requires the root password."

BlueCat™ Linux is an Open Source Linux distribution developed by LynuxWorks[1]. The BlueCat package provides a basic Linux kernel together with tools and utilities that support cross-development work to put Linux in embedded target devices. The commercial distribution of BlueCat supports a range of target architectures including x86, Power PC, ARM and its derivatives, MIPS, and Super H.

LynuxWorks also offers a free downloadable "lite" version of BlueCat that only supports PC-style x86 targets. The book CD includes BlueCat lite in the subdirectory /BlueCat. Because the target is a PC, BlueCat is an ideal tool for learning about, and experimenting with, embedded Linux without investing in a specialized target board.

The "Less is More" Philosophy

There are a number of companies in addition to LynuxWorks offering Linux distributions that target the embedded marketplace. These include Metroworks, RedHat, and Monta Vista[2], to name just a few. Many of these embedded toolkits take the approach of encapsulating the Linux development process inside another environment intended to be easier and more intuitive for an embedded developer new to Linux. This is all well and good to the extent that the vendor's environment fully encapsulates the Linux development process. If there are gaps in the encapsulating

[1] LynuxWorks didn't actually begin with Linux. They've been around for about 14 years providing a hard real-time version of Unix called LynxOS that is now ABI-compatible (Application Binary Interface) with Linux.

[2] Monta Vista is the only one of these vendors that is not accessible over the Net by the obvious web address. Their website is *www.mvista.com*.

environment, you may end up having to cope with substantial elements of Linux development as well as the vendor's tools and environment.

LynuxWorks has taken the opposite approach of simply packaging the standard Linux kernel together with a few simple tools to support the target and cross-development environments. The workflow is basically that of standard Linux but the process is well documented in the BlueCat Users' Guide. Minimizing the modifications to the standard Linux code base makes it easier to track changes in Linux.

Installing BlueCat Linux

The lite version takes around 620 Mbytes of disk space including a 2.6.0 kernel source tree. Follow these steps to install BlueCat under your home directory.

1. Mount the book CD, usually under /mnt/cdrom. If you're running a desktop environment like Gnome or KDE (why wouldn't you?), the book CD will normally be "automounted" as soon as it's inserted.
2. Go to /tmp and create a directory BlueCat
3. In a shell window, cd /tmp/BlueCat
4. tar xzf /mnt/cdrom/bc-5.0-x86-lite.tar.gz
5. Go to your home directory and make a subdirectory BlueCat
6. cd BlueCat.
7. Run the BlueCat install script: /tmp/BlueCat/install

This completes the BlueCat installation procedure. There is an additional tar file of demo code and documentation on the book CD named BlueCatdemo.tar.gz in /mnt/cdrom/BlueCat. Untar BlueCatdemo.tar.gz in BlueCat/.

The commercial version of BlueCat can also be installed on a Windows host if you're so inclined. This uses the CYGWIN execution environment from Cygnus Solutions (now part of Red Hat).

Exploring BlueCat Linux

Take a look at the BlueCat directory (Figure 5-1). With three exceptions all of the subdirectories have names identical to those in the standard filesystem hierarchy. They serve the same purpose as the standard directories except they apply to the *target*. They contain executable tools and configuration files that run on the target. Many of these directories become part of the target by including them in the target file system.

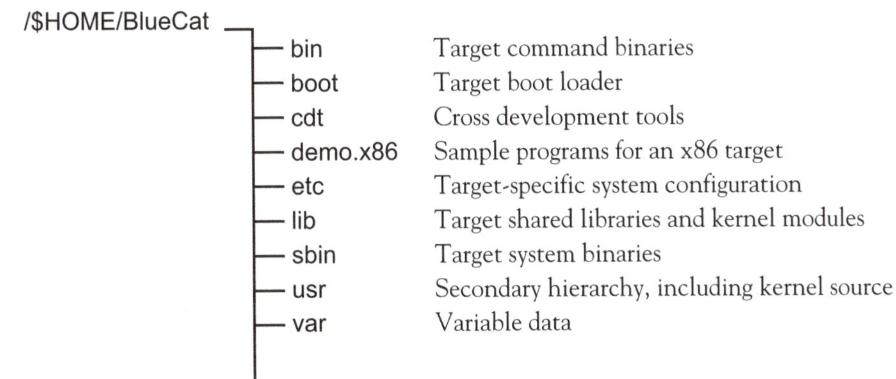

Figure 5-1: BlueCat Linux Directory Structure

The directory cdt/ stands for "cross development tools" and contains the tools necessary to build and debug a target system including, among others, the GNU C compiler and GDB. Under cdt/ you'll find another tree structure that looks suspiciously like the standard filesystem hierarchy. Even though we're building for an x86 target, we have to use the tools in cdt/ rather than the normal set of GNU tools.

usr/src/ has a complete source tree for kernel 2.6.0 modified by LynuxWorks to support their model of embedded development. All BlueCat-specific changes are encapsulated in the source code with #ifdef CONFIG_BLUECAT_xxx or with #ifdef __bluecat__. A search of the .config file will reveal the full set of CONFIG_BLUECAT symbols and then a search of the kernel directories will reveal what files were modified.

The directory demo.x86/ contains several subdirectories with example systems and applications. osloader/ and showcase/ came from the BlueCat distribution while the others came from BlueCatdemo.tar.gz.

The BlueCat Environment

There is a script file in your BlueCat/ directory named SETUP.sh. Its job is to set up the appropriate environment for running the BlueCat tools. It exports several environment variables and puts several paths at the beginning of the PATH environment variable so that the BlueCat tool chain is invoked when you make a project. You must run SETUP.sh before working in the BlueCat environment.

However, you must run SETUP.sh using the "dot built-in" command as follows:

```
. SETUP.sh
```

where there is a space between the dot and the file name. Normally, when Bourne Again Shell (bash) executes a command or script it forks a new process that inherits the exported environment variables of the parent. If we set or change environment variables from the script, the changes only apply to the process executing the script. When it finishes, the changes vanish along with the process.

But bash also has several "built-in" commands, among which is "dot." The dot command runs its script argument in the current process so that the changes take effect in the current process's environment.

Try it out

Try executing SETUP.sh in the normal manner. First, enter the command "set" to see what environment variables are currently defined.

Now run SETUP.sh. The execute permission bit must be set and if your PATH doesn't include the current directory, you'll have to enter ./SETUP.sh. Execute set again. Any changes?

Now use the dot command and enter . SETUP.sh. Run set again.

X86 Target for BlueCat Linux

BlueCat Linux will run small systems quite nicely, even on a 386. This makes it relatively low cost and painless to set up a target for experimenting with BlueCat. Here's your chance to dust off that old 486 box that's been gathering dust in the closet or serving as a door stop and do something useful with it.

Minimal Requirements

We don't need much to get up and running. In principle we don't even need a keyboard or monitor because Linux can boot and run "headless." Nevertheless, it's useful to have a keyboard and display initially to make sure the box is basically running. Here are the minimal requirements for a BlueCat Linux target:

- 386 or higher motherboard with at least 16 Mbytes of RAM.
- Diskette drive, preferably 3.5 inch so it's compatible with your host system. This holds the boot kernel image and root filesystem.

- Serial port.
- Parallel port. We'll use this for experimenting with I/O.
- Network adapter. You can live without it but it really isn't fun. Any common network adapter should suffice. Mine happens to be an NE2000 compatible BOCALANcard 2000 Plus Combo.

That's it. This configuration is roughly equivalent to a small single board computer (SBC) that you might use in a real embedded project. The diskette takes the place of a small flash memory device.

Setup

It's a good idea to have a bootable DOS diskette to do the initial checkout of your target system. If you don't happen to have a bootable diskette, there's an image of one on the book CD named `/BlueCat/W95boot.img`. Use the DOS utility **rawrite** in `/BlueCat` to transfer this image to a diskette. Just start the program and follow the prompts. Note that this process overwrites the entire diskette.

Set up your box with a keyboard and monitor. The first time you power up you should go into BIOS setup, which is usually accomplished by hitting the Del key while the RAM test is in progress. Check that the disk drive configuration matches what you have. I found that POST objected after I removed the hard drive but neglected to change the BIOS setting for Drive C. Exit BIOS setup and boot the DOS diskette. If the boot is successful you've verified most of the motherboard logic and the diskette drive.

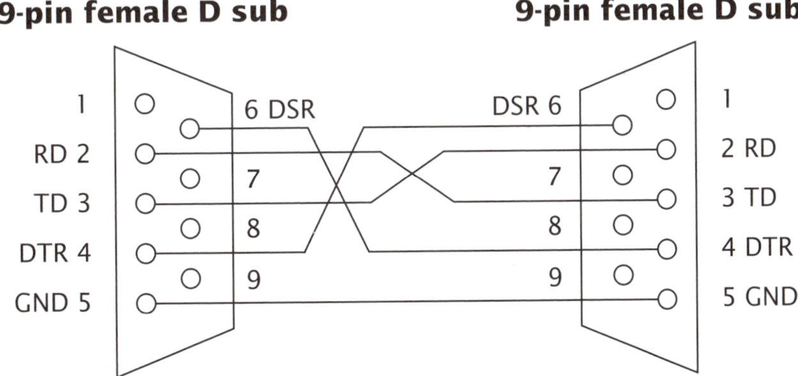

Figure 5-2: Null Modem Cable

To connect the serial port to your workstation, you will most likely need a "null modem." This is a cable with two 9-pin female D sub connectors wired as shown in

Figure 5-2. Transmit Data (TD) and Receive Data (RD) must be swapped so that TD from one end goes to RD on the other end. Likewise the control signals Data Set Ready (DSR) and Data Terminal Ready (DTR) must be swapped. Connect the serial port to COM2 on your workstation.

W95boot.img contains a simple terminal emulator program, term, to check operation of the serial port. term simply echoes back anything it receives on a serial port at 9600 baud and displays it on the target's monitor. Usage is:

```
term <port>
```

where <port> specifies the serial port number, either 1 or 2. <port> is optional and defaults to 1. Type any key to exit the program.

You can exercise the target serial port by running the terminal emulator program minicom from your host workstation. But first, you'll need to configure minicom as described in the next section.

There are two ways to connect the workstation and target to Ethernet. If a network is available, just plug them both into hub ports. Alternatively, you can connect them directly together using a crossover patch cable. Like the null modem, a *crossover* cable swaps the signal pairs so that the transmitter at one end is connected to the receiver at the other end. There's not much we can do about networking at this point. We'll have to wait until we boot a BlueCat image.

Configuring the Workstation

There are some elements on the host workstation that need to be configured in order to communicate with the target. You need to be the root user in order to make most of the changes discussed here. In KDE you can open a file manager window in root (also called Super User) mode. Open the start menu. Go to System Tools->File manager (Super User Mode). Again you'll be prompted for the root password. The new file manager window has root privileges allowing you to edit and change permissions of files and directories owned by root.

The Terminal Emulator, minicom

minicom is a fairly simple Linux application that emulates a dumb RS-232 terminal through a serial port. The default minicom configuration is through a modem device, /dev/modem. We need to change that to talk directly to one of the PC's serial ports.

In a shell window as root user, enter the command minicom –s. If you're running minicom for the first time you may see the following warning message:

WARNING: Configuration file not found. Using defaults.

You will be presented with a configuration menu. Select Serial port setup. Type "A" and replace the word "modem" with either "ttyS0" or "ttyS1", where ttyS0 represents serial port COM1 and ttyS1 represents COM2. Type "E" and type "E" again to select 9600 baud. Figure 5-3 shows the final serial port configuration.

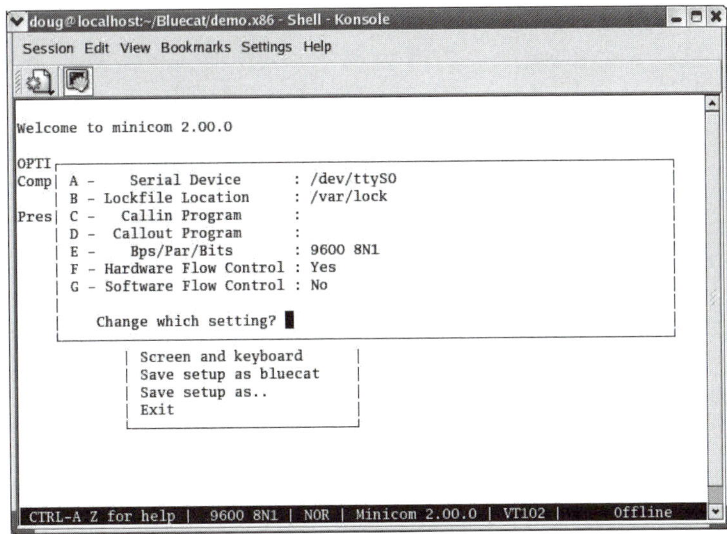

Figure 5-3: minicom Serial Port Settings

Type <Enter> to exit Serial port setup and then select Modem and dialing. Here we want to delete the modem's Init string and Reset string since they're just in the way on a direct serial connection. Type "A" and backspace through the entire Init string. Type "B" and do the same to the Reset string.

Type <Enter> to exit Modem and dialing and then select Screen and keyboard. Type "B" once to change the Backspace key from BS to DEL.

Type <Enter> to exit Screen and keyboard. Select *Save setup as…* Give the configuration a name, "bluecat" for example. Now select Exit from minicom. The next time you start minicom, enter the command minicom <config> where <config> is the name you just gave to your configuration.

You will probably have to change the permissions on the file /dev/ttyS1 to allow the group and world to read and write the device. And of course, you must be root user to do this.

Networking

Your workstation is probably configured to get a network address via Dynamic Host Configuration Protocol (DHCP). But in this case, to keep things simple, we're going to specify fixed IP addresses for both ends.

KDE has a nice graphical menu for changing network parameters. From the Start menu, select System->Network Configuration. In the Network Configuration dialog box, select the Devices tab, select eth0 and click Edit. In the Ethernet Device dialog, select the Protocols tab and TCP/IP, then click edit. In the TCP/IP Settings dialog, unselect the "Automatically obtain IP address settings with..." box. Now enter the fixed IP address. Network address 192.168.1 is a good choice here because it's one of a range of network addresses reserved for local networks. For historical reasons I choose node 11 for my workstation. Also enter the Subnet Mask and Default Gateway Address as shown in Figure 5-4.

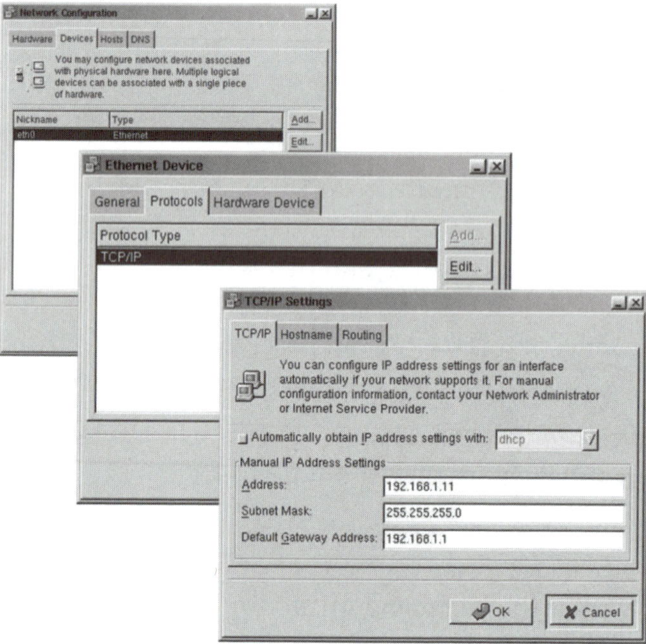

Figure 5-4: Setting Network Parameters

Alternatively you can just go in and directly edit the network device parameters file. Network configuration parameters are found in /etc/sysconfig/network-scripts/ where you should find a file named something like ifcfg-eth0 that contains the parameters for network adapter 0. Make a copy of this file and name it dhcp-ifcfg-eth0. That way you'll have a DHCP configuration file for future use if needed. Now open the original file with an editor (as root user of course). It should look something like Listing 5-1a. Delete the line BOOTPROTO=dhcp and add the four new lines shown in Listing 5-1b. Strictly speaking, the Gateway entry is only necessary if the workstation is connected to a network with Internet access.

```
DEVICE=eth0
ONBOOT=yes
BOOTPROTO=dhcp
```

Listing 5-1a: ifcfg-eth0

```
DEVICE=eth0
ONBOOT=yes
IPADDR=192.168.1.11
NETMASK=255.255.255.0
GATEWAY=192.168.1.1
BROADCAST=192.168.1.255
```

Listing 5-1b: revised ifcfg-eth0

We'll use NFS (network file system) to download images to our target. That means we have to "export" one or more directories on the workstation that the target can mount on its filesystem. Exported directories are specified in the file/etc/exports. Initially this file is present but empty. Open it with an editor and insert the following lines:

```
/home/<your_home_name>/BlueCat

/home 192.168.1.200 (r, no_root_squash)
```

We'll deal with the first line in the next chapter. The second line represents the location from which the target will try to download a kernel image and a root filesystem. The reason for choosing /home will become apparent later. The IP address is that of the target. Again, for historical reasons I choose 200 as the node address for the target.

Finally, we have to start the NFS server. This is accomplished with the command /etc/rc.d/init.d/nfs start. You can execute this command from a shell window or, better yet, add it near the end of /etc/rc.d/rc.local. This is the last script executed at boot up.

First Test Program

We're finally ready to download and run something on our target. The best place to start is the osloader demo. This will show how demo systems are organized and built, and also give us a boot loader that can boot other kernel images over the network.

A demo directory roadmap

All of the demo directories are organized with essentially the same files and subdirectories. Take a look at /$HOME/BlueCat/demo.x86/osloader as an example. In the following descriptions <project> is the name of the project and the subdirectory under demo.x86/. The project directory breaks down as follows:

Directories

local/ — Configuration files specific to this project. These generally end up in /etc or its subdirectories on the target.

src/ — Source files of applications or kernel extensions specific to this project.

Configuration Files

<project>.config — Config file for this project's kernel. When you do **make xconfig** this file is copied to **.config** in the kernel source tree. The modifications are made there and then the file is copied back to <project>.config. when you execute **make kernel** this file is again copied to **.config** in the kernel source tree.

<project>.spec — This file provides information to build the root file system. The syntax looks like shell commands to create directories and nodes and copy files. It is interpreted by the BlueCat utility **mkrootfs**.

cl.txt — I added this one. It contains a command line argument for the kernel to redirect the console to ttyS0.

Makefile

The project Makefile supports a number of targets as follows:

xconfig — Copies <project>.config to usr/src/linux/.config and invokes **make xconfig** from there. Then it copies the resulting .config file back to <project>.config.

kernel	Builds the kernel based on <project>.config. Again, <project>.config is copied to usr/src/linux/.config and the normal kernel Makefile is invoked from there.
rootfs	Builds the root file system based on <project>.spec.
kdi	Builds the <project>.kdi target. See below.
this	Builds the custom programs in src/.
all	Builds all of the above targets except xconfig.
clean	Removes all of the target intermediate files.

Target Files

These files are the result of building the Makefile targets:

<project>.kernel	Compressed kernel image suitable for booting onto a target board over a network using the BlueCat Linux OS loader. Built by make kernel.
<project>.disk	Compressed kernel image suitable for copying onto a floppy or hard disk. Built by make kernel.
<project>.rfs	Compressed RAM disk root file system image suitable for booting onto a target board over a net-work using the BlueCat Linux OS loader, or for loading from a floppy disk or hard disk. Built by make rootfs.
<project>.tar	Tar image of the root file system suitable for copying on a hard disk partition or for NFS-mounting. Built by make rootfs.
<project>.kdi	This target contains the compressed kernel image (.disk) and the compressed RAM disk root file system (.rfs). It is suitable for booting onto a target board from a network using firmware, or programming into ROM/flash memory on the target. Built by make kdi.

Making a Boot Disk

osloader/ already has a kernel image and root filesystem that should run fine on your target unless you have a truly bizarre network interface. The reason for choosing /home as the location from which to boot a secondary kernel image and root filesystem is that every Linux system has a /home directory. This way you can use the supplied root filesystem image without having to modify any configuration parameters to point to someplace in your home directory.

Now it's time to create a bootable diskette. For that we use several invocations of the BlueCat utility **mkboot**. With a clean, or otherwise unnecessary, diskette in the workstation drive, execute the following set of commands in the directory /$HOME/BlueCat/demo.x86/osloader:

mkboot –b /dev/fd0	Writes a boot sector to the diskette.
mkboot –k osloader.disk /dev/fd0	Copies the kernel image to the diskette.
mkboot –c cl.txt /dev/fd0	Creates a command line for the kernel using the contents of cl.txt.
mkboot –f osloader.rfs /dev/fd0	Copies the root filesystem to the diskette.
mkboot –r /dev/fd0 /dev/fd0	Sets the device node on the target board to mount as the root file system or uncompress the file system image.

After each of these operations **mkboot** will report on the current status of the diskette. The first time you build a bootable diskette it is important that you do these steps in the order shown. Later on you may find it necessary to modify one or more elements of the root filesystem. You may then execute the **mkboot –f** command without executing any of the commands that came before it.

However if you change the kernel and have to rebuild it, you will have to re-execute the **mkboot –k** command. More than likely the new kernel is larger than the old one and this will overwrite part of the root filesystem. You will then have to re-execute the **mkboot –f** command.

Executing the Target Image

Take the diskette created in the previous section and insert it into your target system. Turn on power or reboot and while the target is coming up, execute **minicom bluecat** (or whatever name you saved the minicom configuration under) in a shell window on the workstation. You should see the normal stream of messages that come out of a Linux kernel booting up.

This particular system doesn't do a whole lot. About all it can do in fact is to boot another kernel image and root filesystem through NFS using BLOSH, the Bluecat Linux lOader SHell. BLOSH is a shell-like utility with a small set of built-in commands for downloading and executing file images.

When the target finishes booting it presents the BLOSH prompt, ">". Type "set" to see the BLOSH environment variables. These define the target environment parameters such as network interface and IP address, NFS host, ker-nel image and root filesystem (see Figure 5-5). BLOSH gets these values from a startup file called **blosh.rc** in osloader/local. If you want to change the location from which BLOSH boots a kernel and root filesystem over NFS, edit **blosh.rc** and rebuild the osloader root filesystem.

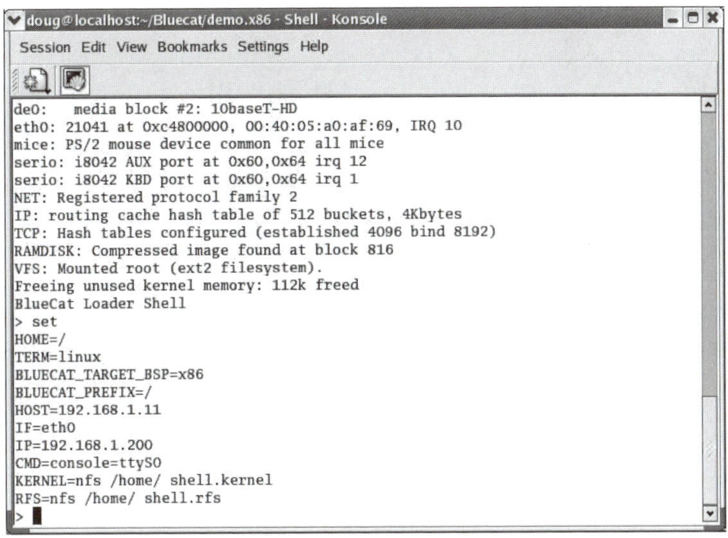

Figure 5-5: BLOSH Environment Variables

The "Shell" Kernel

To proceed any further, we need a kernel with NFS support and a root filesystem with some minimal set of shell utilities. This is in the project /opt/BlueCat/demo.x86/shell/ on the workstation. Edit shell/local/rc.sysinit. The last line mounts an NFS volume on the local filesystem. Replace "doug" with your home directory name. rc.sysinit is the script file normally interpreted by the init process when the Linux kernel starts.

In shell/, execute make rootfs to build a new root filesystem containing the revised rc.sysinit. Note that shell.kernel already exists and should work for your target. Now copy the files shell.kernel and shell.rfs to /home. Note in Figure 5-5 that these are the files that BLOSH expects to download as the kernel image and root file-system, respectively.

Back in the shell window with minicom running, enter the command boot. This will cause the files shell.kernel and shell.rfs to be downloaded through NFS. The Linux kernel contained in shell.kernel will then start and it will mount the root filesystem contained in shell.rfs.

When the boot command completes (possibly including some timeout errors that appear to be innocuous), you will be presented with a standard bash prompt. Try out some simple commands like ls. cd to /proc and have a look at some of the /proc files using the cat command.

Finally, take a look at the directory /usr. An ls command executed on this directory will show that it is in fact the same as /$HOME/BlueCat on the workstation. Hmmm… How did that happen? Well, if you remember, earlier in this chapter we exported the directory /home/<your_name>/BlueCat through NFS.

The line you edited in rc.sysinit is a "mount" command that uses NFS to mount the directory /home/<your_name>/BlueCat on the workstation as /usr on the target system. So now when we reference a file in /usr on the target, it really references a file on /$HOME/BlueCat on the workstation.

Echo the PATH environment variable on the target. Note that it includes /usr/bin and /usr/sbin. These now map directly to bin/ and sbin/ in BlueCat/ on the workstation. This means we can execute target programs *stored on the workstation's disk.* We'll exploit this capability to definite advantage in the next chapter.

Resources

There is a reasonably good manual that accompanies BlueCat Linux. It describes the various utilities that are provided to support the embedded target environment. Oddly enough it is not part of the downloadable BlueCat lite distribution. Instead it is included in the BlueCatdemo.tar.gz file that you untarred at the beginning of the chapter. So you'll find it in /$HOME/BlueCat/. The release notes are also available here.

For greater insight into what happens as the kernel boots up, check out these HOWTOs at The Linux Documentation Project:

PowerUp-to-Bash-Prompt-HOWTO
Linux-Init-HOWTO
Kernel-HOWTO

CHAPTER 6

Debugging Embedded Software

"Use the Source, Luke, use the Source. Be one with the code. Think of Luke Skywalker discarding the automatic firing system when closing on the deathstar, and firing the proton torpedo (or whatever) manually. **Then** *do you have the right mindset for fixing kernel bugs."*

—Linus Torvalds

The Target Setup

Before diving into the subject of debugging, let's review our setup so far. We have a minimal kernel and minimal root filesystem on diskette that allows us to boot other kernel images and root filesystems over the network via NFS. stdin, stdout, and stderr on the target are connected to ttyS0, which in turn is physically connected to ttyS0 or ttyS1 on the host. We communicate with the shell running on the target through the terminal emulation program minicom. The boot loader script directs it to boot a kernel image and root filesystem named shell.

The shell kernel is configured to start up a bash shell with a minimal set of command utilities in /bin and /sbin. It also mounts an NFS volume on /usr. The mounted volume is $HOME/BlueCat on the host workstation. Consequently every file in the BlueCat/ directory is accessible from the target. The implication of this, among other things, is that we can execute *on the target* program files that physically reside on the host's filesystem. This allows us to test user space application programs without having to build and download a new root filesystem. And of course, programs running on the target can open files on the NFS-mounted host volume. Again, the console device is redirected to ttyS0 via a command line option to the kernel so that we can communicate with bash running on the target. Figure 6-1 shows graphically how these elements interrelate.

In principle we could even use the network to communicate with the shell. We simply open a telnet session on the host directed at the target. But BlueCat™ lite doesn't include a telnet daemon so we'll just stick with what we have for now.

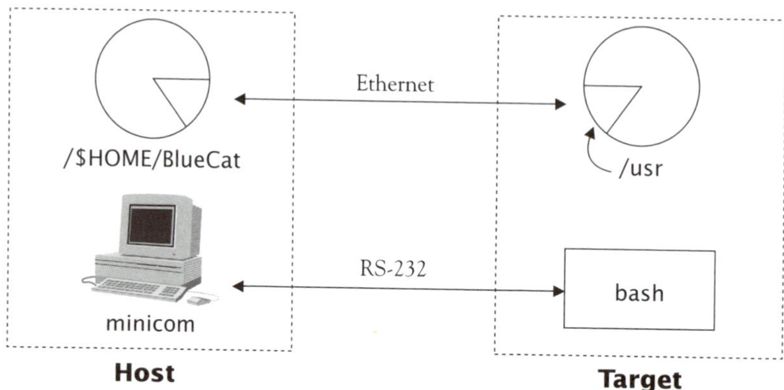

Figure 6-1: The Host and the Target

It would be nice if we could put our shell kernel and root filesystem on a diskette like the osloader. Unfortunately, `shell.kernel` plus `shell.rfs` adds up to more than the 1.44 MByte capacity of a diskette. A genuine target board would have some form of flash memory, perhaps 4 or 8 MBytes, enough to hold a kernel image and a reasonable root filesystem. So, for now anyway, we're stuck with booting shell over NFS.

GDB

GDB stands for the GNU DeBugger, the source-level debugging package that is part of the GNU toolchain. Like any good debugger, it lets you start your program, insert breakpoints to stop the program at specified points, examine and modify variables and processor registers. Like most Linux utilities, it is command-line driven making it rather tedious to use. The solution is to front GDB with a graphical front end such as a data display debugger, DDD. This combination makes a truly usable source level debugger.

In a typical desktop environment, the target program runs on the same machine as the debugger. But in our embedded environment DDD/GDB runs on the host and the program being debugged runs on the target (see Figure 6-2). Fortunately, GDB implements a serial protocol that allows it to work with remote targets either over an RS-232 link or Ethernet.

There are two approaches to interfacing the target to the GDB serial protocol:

- *gdb stubs*. A set of functions linked to the target program. gdb stubs is an RS-232-only solution.
- `gdbserver`. This is a stand-alone program running on the target that, in turn, runs the program to be debugged. The advantage to `gdbserver` is that it is

totally independent of the target program. In other words, the target builds the same regardless of remote debugging. The other major advantage of gdbserver is that it runs over Ethernet.

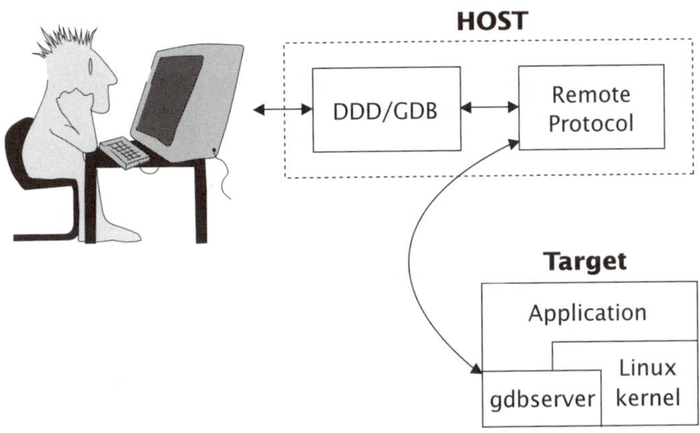

Figure 6-2: Remote Debugging with gdb

DDD Front End

DDD provides an X-windows front end for GDB. Thus you can point-and-click to set breakpoints, examine variables and so on. DDD translates the graphical user interface (GUI) input into commands for GDB, known in this environment as the "inferior" debugger.

Most recent Linux distributions include DDD and of course GDB is part of the standard GNU toolchain. The book CD includes gdbserver and, if needed, DDD under the directory /tools. Copy gdbserver to BlueCat/bin.

Like most Open Source code, DDD is distributed in source form. If needed, untar ddd-3.3.1.tar.gz into your home directory. cd to ddd-3.3.1/, read the INSTALL file, and follow the directions.

tools/ also has a copy of the extensive, reasonably well-written users manual for DDD in PDF format.

Debugging a Sample Program

The easiest way to get started with a new software tool is to just try it out. So before getting into the details of remote debugging, let's just try out DDD on the host and play with it a bit. In demo.x86/apps/ is a subdirectory called ddd/. In there is a small program called shellsort.

Open Makefile with an editor. This is a trivial make file and it might be almost as easy to type the compile command itself. But Makefile will turn out to be useful later on from within DDD. The –g option in the compile command adds symbol information to the executable for use by GDB.

Build the program with the command:

```
make
```

This program sorts and prints out in ascending order the list of numeric arguments passed to it. For example:

```
./shellsort 4 6 3 1 7 8

1 3 4 6 7 8
```

Most of the time it works. But there's a subtle bug in the program. Try for example:

```
./shellsort 4000 1000 7000 6000 8000
```

The arguments are sorted in the correct order but one of them has been changed. Let's use DDD to find out what's happening. Enter the command:

```
ddd shellsort
```

After a few seconds the display shown in Figure 6-3 appears. The various elements of this display are:

Source Window: Displays the source code around the current execution point.

Command Tool: Buttons for commonly used commands. This is a separate window that can be moved around within the source window.

ToolBar: Contains buttons for commands that require an argument along with a window for entering the argument.

Debugger Console: Lets you enter commands directly to the inferior debugger's command line. There are some operations that just work better at the command line.

Status Line: Shows the current state of DDD and GDB.

We'll encounter other windows as we go along.

The first thing to do now is to set a Breakpoint so that the program will stop at a location we are interested in. Click on the blank space to the left of the line that initializes a. The Argument window in the Toolbar, identified by ():, now contains the location ('shellsort.c:32') of the cursor. Now click on the Break icon in the Toolbar to create a breakpoint at the location in ():. A little red stop sign appears in line 32. Alternatively you can right click on the left side of line 32. A menu appears where one of the options is Set breakpoint. Selecting this is identical to clicking the Break button in the tool bar.

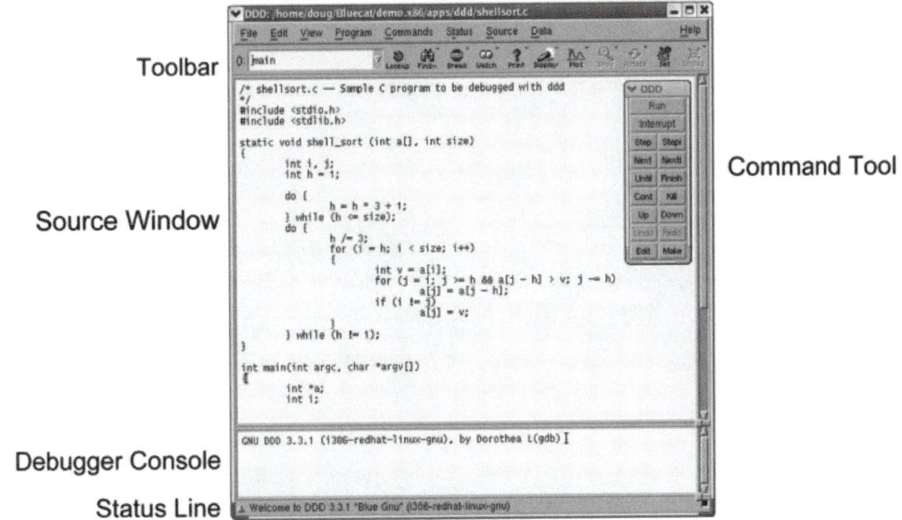

Figure 6-3: DDD Main Window

Now select Program->Run from the menu bar. This brings up the Run Program dialog box. In the Run With Arguments window, enter a set of numbers like the ones above that will cause the program to fail. Click on Run. Very shortly the program halts at the breakpoint in line 32. The Debugger Console reports details of the breakpoint and a green arrow appears at line 32 to show the current execution point.

To see the current value of a variable, just place the cursor over it. In about a second a yellow box appears with the current value. Try it with argc. a hasn't been initialized yet so its value is random. It's a local variable allocated on the stack, so its value is whatever happened to be in that location when the program started. To execute the current line, click on the Next button on the Command Tool. The arrow advances to the following line. Now point again to a to see that the value has changed and has in fact been initialized.

Move the cursor to the left of the call to **shell_sort()** and right click. Select **Set Temporary Breakpoint** and click **Cont** on the Command Tool. The breakpoint we set earlier on line 32 is called a "sticky" breakpoint because it stays there until we specifically remove it. A temporary breakpoint goes away the first time it is hit. Some debuggers call this function "Execute to here."

When the program reaches line 36, the **a** array has been initialized. To view all values assigned to **a**, enter the following in the Argument window on the Toolbar:

```
a[0]@(argc-1)
```

and click the **Print** button in the Toolbar. The values of all five elements appear in the Debugger Console. Because **a** was dynamically allocated, GDB doesn't know how big it is. The "@" notation tells GDB how many elements to list.

Rather than using **Print** at each stop to see the current value of **a**, you can also **Display a**, such that it is automatically updated and displayed each time the program stops. With **a[0]@(argc - 1)** still showing in the Argument window click on **Display** in the Toolbar. The contents of **a** are now shown in a new window, the **Data** window (Figure 6-4). Initially the array is displayed vertically so that most of it is hidden. Click on **Rotate** to rotate the array horizontally making it all visible.

Data Window

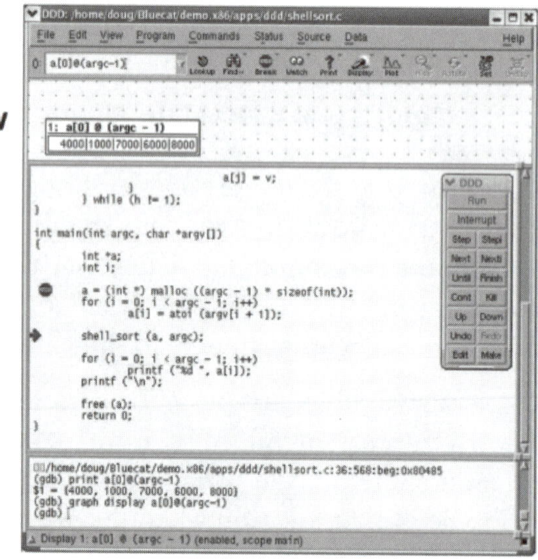

Figure 6-4: DDD Data Window

You can display as many variables as you'd like in the Data window. As you open new variables, they tend to be displayed vertically so that they're not visible without scrolling. You can drag the individual variable windows within the Data window to make visible whatever is most important at the moment.

Click on Step in the Command Tool to step into the shell_sort() function. The execution point arrow moves to the first executable line of shell_sort() and the Debugger Console shows the arguments passed to it. Click on Status->Backtrace in the menu. This brings up a "stack frame" display shown in Figure 6-5. It shows the specific series of calls that led you to this function. You can to see where you are in the stack as a whole. Selecting a line (or clicking on Up or Down) will let you move through the stack.

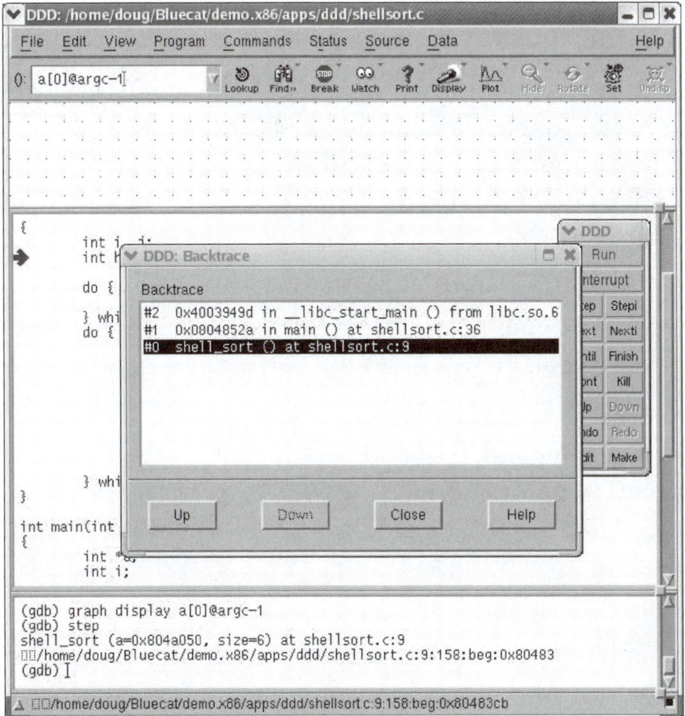

Figure 6-5: DDD Backtrace Display

Note how the a display disappears when main()'s stack frame is left. That's because a is allocated locally within main(), on the stack, and so is not visible, i.e., is "out of scope," in the shell_sort() function.

There's an important distinction between Next and Step on the Command Tool. If the cursor is sitting on a function call, shell_sort() for example, clicking Next steps *over* the function call, clicking Step steps *into* it.

Let's check whether the arguments to shell_sort() are correct. After returning to the lowest frame, enter a[0]@size in the argument field and click on Print:

```
(gdb) print a[0] @ size
$2 = {4000, 1000, 7000, 6000, 8000, 4001}
(gdb)
```

Aha! Where did 4001 come from? We told GDB to print size elements and it printed six. But we only entered five arguments to shellsort. So the size argument being passed to shell_sort() is off by one. Note incidentally that you may see a different value for that last element of a. This is what happens to show up on my system.

To verify that the size argument is wrong, we can change its value and see what happens. Select size in the source code and click on Set on the right hand side of the Toolbar. A dialog pops up where you can edit the variable value. Change the value to 5 and click OK. Then click Finish on the Toolbar to let the program run to completion. Now the result is correct.

If you look back at the call to shell_sort() in main(), you see that the size argument is argc. Oops! argc counts the name of the program as the first argument. And in fact we saw earlier that argc was 6. So the correct size argument to shell_sort() should be argc – 1.

Click on Edit in the Command Tool and voila!, up pops the source file in vi centered at the current execution point. Correct the parameter in the shell_sort() call and click on Make to rebuild the program. Now select Run->Run Again from the menu. The program starts over executing the corrected code and the revised code shows up in the Source window.

The reason this simple project uses a make file instead of just typing the single compile command is that the Make button requires it.

If vi doesn't happen to be your favorite editor, you can define the environment variable XEDITOR to set the editor DDD uses. For C files, I like the KDE advanced editor kate with syntax coloring. So I added:

```
XEDITOR=kate
```

to my .bash_profile.

Note that most of the icons in the Toolbar have a little down arrow in the upper right hand corner. If you click the icon and hold the left mouse button down, a drop-down menu appears that modifies the behavior of the basic icon. For example, the Display icon gives you the option of converting a variable to any of several bases.

Not surprisingly, for a program as capable as DDD, there are a great many settable and configurable options. Select Edit->Preferences and Edit->GDB Settings from the menu to get a feel for the many parameters that can be configured. The accompanying Help text is reasonably helpful.

Clearly this brief tour of DDD has only touched on the highlights. Yet the few commands we've looked at thus far are sufficient for most of your debugging needs.

Setting up for Remote Debugging

Our next step is to move the shellsort program to the target and debug it there.

gdbserver

gdbserver resides in /$HOME/BlueCat/bin, which translates to /usr/bin on the target and happens to be in the target's PATH. In the target window, the one running minicom, cd /usr/demo.x86/apps/ddd. Start gdbserver with the command:

```
gdbserver :1234 shellsort 4000 1000 7000 6000 8000
```

The arguments to gdbserver are:

- Network port number preceded by a colon. The normal format is "host:port," where "host" is either a host name or an IP address in the form of a dotted quad, but gdbserver ignores the host portion. The port number is arbitrary as long as it doesn't conflict with another port number in use. Generally, port numbers below 1024 are reserved for established network protocols, so it's best to pick a number above 1023.
- Target program.
- Arguments to the target program.

gdbserver responds with:

```
Process shellsort created; pid = 22
```

Your pid value may be different. This says that gdbserver has loaded the program and is waiting for a debug session to begin on the specified port.

Connect GDB

Start up DDD as above. Now before starting execution, we have to connect GDB to the target. In the Debugger Console window, enter:

```
target remote 192.168.1.200:1234
```

This tells GDB to connect to a remote target, more specifically a network target since the target address is in the form "host":"node." Here the "host" portion is significant as it identifies the target's node address. The "port" number must match the one used when starting gdbserver on the target.

Observe the output from gdbserver showing that the target program has started. Note the corresponding message in the Debugger Console window.

When connected to a remote target, GDB thinks the target program has already started. Set up an initial breakpoint as before, but now use the Cont button on the Command Tool instead of Run to proceed. Run through the steps we did above on the workstation to verify that GDB behaves the same.

When shellsort terminates on the target, a message appears in the Debugger Console showing the program's return value, in this case 0. Note however, that the target has not presented another bash prompt. That's because the program we started on the target is gdbserver and not shellsort. gdbserver is still running even though shellsort has terminated. gdbserver terminates when you close ddd on the workstation.

Note incidentally that both gdbserver and shellsort are run from the NFS mounted volume on the host.

The Host as a Debug Environment

Although remote GDB gives us a pretty good window into the behavior of our program on the target, there are still reasons why it might be useful to do initial debugging on our host development machine. To begin with, the host is available as soon as the project starts, probably well before any real target hardware is available or working. The host has a file system that can be used to create test scripts and document test results.

Of course, in order to use this technique, you must have both target and host versions of your operating system. In our current BlueCat environment that's not a problem since both the host and target are x86. But even if the target is different, with Linux it's not a problem since we have the source code by definition.

When you characterize the content of most embedded system software, you will usually find that something like 5% of the code deals directly with the hardware. The rest of the code is independent of the hardware and therefore shouldn't need hardware to test it, provided that you can supply the appropriate stimulus to exercise it.

Unfortunately, all too often the 5% of hardware-dependent code is scattered randomly throughout the entire system. In this case you're virtually forced to use the target for testing because it is so difficult to "abstract out" the hardware dependencies. The key to effective host-level testing is rigorous software structure. Specifically, you need to isolate the hardware-dependent code behind a carefully defined API that deals with the hardware at a higher, more abstract level. This is not hard to do but it does require some planning.

If your hardware-dependent code is carefully isolated behind a well-defined API and confined to one or two source code modules, you can substitute a *simulation* of the hardware-dependent code that uses the host's keyboard and screen for I/O. In effect, the simulated driver "fools" the application into thinking the hardware is really there.

You can now exercise the application by providing stimulus through the keyboard and noting the output on the screen. In later stages of testing you may want to substitute a file-based "script driver" for the screen and keyboard to create reproducible test cases.

Think of this as a "test scaffold" that allows you to exercise the application code in a well-behaved, controlled environment. Among other things, you can simulate limit conditions that might be very difficult to create, and even harder to reproduce, on the target hardware.

The "Thermostat" Example

While the `shellsort` program might be interesting, it has little to do with real embedded system problems. Here is a sample program closer to the real world that we'll use for the remainder of this chapter and on and off throughout the rest of the book.

`cd /$HOME/BlueCat/demo.x86/apps/thermostat` and open the file `thermostat.c` with your favorite editor. This is a simple implementation of a thermostat. If the measured temperature drops below a specified setpoint, a "heater" turns on. When the temperature rises above the setpoint the heater turns off. In practice, real thermostats incorporate hysteresis that prevents the heater from rapidly switching on and off when the temperature is right at the setpoint. This is implemented in the form of a "deadband" such that the heater turns on when the temperature drops below the setpoint – deadband and doesn't turn off until the temperature reaches setpoint + deadband. Additionally, the program includes an "alarm" that flashes if the temperature exceeds a specified limit.

thermostat.c is fundamentally a state machine that manipulates two digital outputs in response to an analog input and the current state of the program. Note in particular that thermostat.c makes no direct reference to any I/O device. The analog input and digital outputs are virtualized through a set of "device driver" functions. This allows us to have one set of driver functions that works in the host simulation environment (simdrive.c) and another set that works on the target (trgdrive.c).

For our purposes the driver API (Application Programming Interface) is fairly simple and includes the following functions:

- int initAD (void). Does whatever is necessary to initialize the A/D converter. If the initialization is not successful, it returns a non zero status value.
- unsigned int readAD (unsigned int channel). Returns the current value for the channel that was passed in the previous call to readAD() and sets the analog multiplexer to *channel*. This allows for a multichannel implementation although it's not used here.
- void closeAD (void). Does whatever is necessary to "clean up" when the program terminates. May not be necessary in some implementations.
- setDigOut (int bitmask). Turns on the bit(s) specified by bitmask.
- clearDigOut (int bitmask). Turns off the bit(s) specified by bitmask.
- For completeness, we should probably have a pair of functions like initDigOut() and closeDigOut(). In this particular implementation they simply turn out to be unnecessary.

Take a look at simdrive.c. This driver implementation uses a shared memory region to communicate with another process that displays digital outputs on the screen and accepts analog inputs via the keyboard. This functionality is implemented in devices.c. The shared memory region consists of a data structure of type shmem_t (defined in driver.h) that includes fields for an analog input and a set of digital outputs that are assumed connected to LEDs. It also includes a process ID field (pid_t) set to the pid of the devices process that allows the thermostat process to signal when a digital output has changed.

devices creates and initializes the shared memory region. In simdrive, initAD() attaches to the previously created shared memory region. readAD() simply returns the current value of the a2d field in shared memory. The setDigOut() and clearDigOut() functions modify the leds field appropriately and then signal the devices process to update the screen display. Figure 6-6 illustrates the process graphically.

Build both devices and thermostat with the command make sim. To run this example on the host, start up two shell windows. In the first, run devices and in the second run thermostat.s (the .s identifies the simulation version). Change the "A/D in:" input in the devices window and verify that the thermostat responds correctly. Run thermostat.s under DDD to see how the state transitions progress.

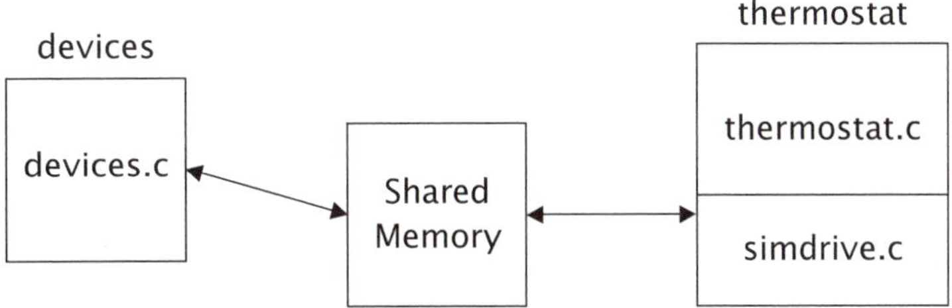

Figure 6-6: Thermostat Simulation

Adding Programmable Setpoint and Limit

As presently designed, the setpoint, limit, and deadband for the thermostat are "hardwired" into the program, which is probably not particularly useful. At the very least, setpoint and limit should be adjustable, preferably remotely. The obvious solution is to use stdin to send simple commands to change these parameters.

If we were building this under DOS, we would probably use the function kbhit() to poll for console input within the main processing loop in thermostat. But Linux is a multitasking system. Polling is tacky! The proper solution is to create an independent thread of execution whose sole job is to wait for console input and then interpret it.

So here's your chance to do some real programming. Modify thermostat.c to incorporate settable setpoint and limit parameters. We'll use a very simple protocol to set parameters consisting of a letter followed by a number. "s" represents the setpoint and "l" the limit. So to change the setpoint you would enter, for example,

```
s 65<Enter>
```

This sets the setpoint to 65°.

We'll create a new child process that waits for keyboard input, parses the received string and updates the specified parameter. Refer back to the description of fork() in Chapter 2. Add the fork() code to thermostat.c after the call to initAD() and before the while loop. The child process is shown in pseudocode form in Listing 6-1.

Remember though that the child process inherits a *copy* of the parent's data space, not the same data space. So even if the child resides in the same source file as the parent, it sees a different copy of setpoint and limit. Changes made to these variables by the child are not seen by the parent.

The solution is to put the parameters into a data structure and allocate a shared memory region for that structure just as simdrive does for the simulated peripherals. The parent will create the shared memory region and both the parent and child must get and attach to it.

```
char *whitepace = " \t,";

get and attach shared memory region (see simdrive.c for details)

while (1)
{
    char string[20];
    char *token;

    gets (string);
    token = strtok (string, whitespace);
    value = atoi (strtok (NULL, whitespace));

    switch (token[0]);
    {
        case 's':
            // update setpoint
        case 'l':
            // update limit
        default:
            // ignore
    }
}
```

Listing 6-1: Command Parsing Process

If you're not familiar with any of the functions shown here, refer to the corresponding Section 3 main page.

Once you have the program built, try it out under DDD. Note that you'll be using the same console device to enter the command string that thermostat uses to write out the current temperature. This results in a certain amount of unavoidable interference.

Once you're satisfied with the program's operation in the host, the next step is to build it for the target. We'll do that in the next chapter.

There is another, actually somewhat simpler, approach to creating a second thread of execution using Posix Threads. Whereas a *process* carries with it a complete execution context and full set of protected resources, a thread is a thread of execution only. The only resources a *thread* owns are code and a stack.

The thread API is actually more like traditional multitasking programming than the Linux fork() model. That is, you create a thread out of a function by calling a thread creation service. Creating a thread is generally a lower overhead operation than creating a process and threads are often characterized as "light-weight multitasking."

The subject of Posix Threads is covered later in Chapter 12.

Resources

http://www.gnu.org/software. This is the "go to" place for the latest information on Open Source packages maintained by the Free Software Foundation. The quickest way to get to the page for a specific package is to know the package name and append it to the link listed above. So DDD is at *www.gnu.org/software/ddd* and GDB is at *www.gnu.org/software/gdb*. You'll find the latest releases as well as documentation and other resources.

In addition to GDB and DDD, this chapter introduced several Linux programming concepts. An excellent guide to Unix/Linux programming is

Matthew, Niel and Richard Stone, *Beginning Linux Programming 3rd Edition*, Wrox Press, 2003. I happen to have the 2nd edition.

CHAPTER 7

Kernel Modules and Device Drivers

"If you think I'm a bad driver, you should see me putt."
—Snoopy

For me, the fun part of embedded programming is seeing the computer interact with its physical environment, i.e., actually do something. In this chapter we'll use the parallel port on our target machine to provide some limited I/O capability for the thermostat program developed in the previous chapter. But first some background.

Kernel Modules

Installable kernel modules offer a very useful way to extend the functionality of the basic Linux kernel and add new features without having to rebuild the kernel. A key feature is that modules may be dynamically loaded when their functionality is required and subsequently unloaded when no longer needed. Modules are particularly useful for things like device drivers and `/proc` files.

Given the range and diversity of hardware that Linux supports, it would be impractical to build a kernel image that included all of the possible device drivers. Instead the kernel includes only drivers for boot devices and other common hardware such as serial and parallel ports, IDE and SCSI drives, and so on. Other devices are supported as loadable modules and only the modules needed in a given system are actually loaded.

Loadable modules are unlikely to be used in a production embedded environment because we know in advance exactly what hardware the system must support and so we simply build that support into the kernel image. Nevertheless, modules are still useful when testing a new driver. You don't need to build a new kernel image every time you find and fix a problem in your driver code. Just load it as a module and try it.

Keep in mind that modules execute in Kernel Space at Privilege Level 0 and thus are capable of bringing down the entire system.

A Module Example

cd /$HOME/BlueCat/demo.x86/apps/module and open the file hello.c. Note first of all the two include files required for a kernel module. This is a trivial example of a loadable kernel module. It contains two very simple functions; hello_init() and hello_exit(). The last two lines:

```
module_init(hello_init);
module_exit(hello_exit);
```

identify these functions to the module loader and unloader utilities. Every module must include an init function and an exit function. The function specified to the module_init() macro, in this case hello_init(), is called by insmod, the shell command that installs a module. The function specified to module_exit(), hello_exit(), is called by rmmod, the command that removes a module.

In this example both functions simply print a message on the console using printk, the kernel equivalent of printf. C library functions like printf are intended to run from user space making use of operating system features like redirection. These facilities aren't available to kernel code. Rather than writing directly to the console, printk writes to a circular buffer and then wakes up the klogd process to deal with the message by either printing it to the console and/or writing it to the system log.

Note the KERN_ALERT at the beginning of the printk format strings. This is the symbolic representation of a *loglevel* that determines whether or not the message appears on the console. Loglevels range from 0 to 7 with lower numbers having higher priority. If the loglevel is numerically less than the kernel integer variable console_loglevel then the message appears on the console. In any case, printk messages always show up in the file /var/log/messages regardless of the loglevel.

The hello example also shows how to specify module parameters, local variables whose values can be entered on the command line that loads the module. This is done with the module_param() macro. The first argument to module_param() is a variable name, the second is the variable type, and the third is a "permission flag" that controls access to the representation of the module parameter in *sysfs*, a new feature of the 2.6 kernel that offers cleaner access to driver internals than the /proc filesystem. A safe value for now is S_IRUGO meaning the value is read-only.

Variable types can be charp—pointer to a character string, or int, short, long—various size integers or their unsigned equivalent with "u" prepended.

Make hello with a simple make command and then try it out. As root user, enter the command:

```
insmod hello.ko my_string="name" my_int=47
```

If you're running from the command line, i.e., not in X windows, you should see the message printed by hello_init(). Oddly enough, printk messages do not appear in shell windows running under X windows, KDE for example, regardless of the loglevel. You can see what printk did with the command:

```
tail /var/log/messages
```

This prints out the last few lines of the messages log file. A useful variation on this command is:

```
tail –f /var/log/messages
```

This version continues running and outputs new text sent to messages.

If you initially logged on under your normal user name and used su to become root, you'll have to preface all of the module commands with /sbin/ because /sbin is not normally part of the path for a normal user. Now try the command lsmod. This lists the currently loaded modules in the reverse order in which they were loaded. hello should be the first entry. lsmod also gives a "usage count" for each module and shows what modules depend on other modules. The same information is available in /proc/modules.

Now execute the command rmmod hello. You should see the message printed by hello_exit(). Finally execute lsmod again to see that hello is no longer listed. Note incidentally that rmmod does not require the ".ko" extension. Once loaded, modules are identified by their base name.

A module may, and usually does, contain references to external symbols such as printk. How do these external references get resolved? insmod resolves them against the kernel's symbol table, which is loaded in memory as part of the kernel boot process. Furthermore, any nonstatic symbols defined in the module are added to the kernel symbol table and are available for use by subsequently loaded modules. So the only external symbols a module can reference are those built into the kernel image or previously loaded modules. The kernel symbol table is available in /proc/ksyms.

Resources

The subject of module and device driver programming is way more extensive than we've been able to cover here. Hopefully, this introduction has piqued your interest and you'll want to pursue the topic further. An excellent book on the topic is:

Rubini, Alessandro, Jonathan Corbet, and Greg Kroah-Hartman, *Linux Device Drivers, 3nd Ed.*, O'Reilly, 2005.

In fact, I would go so far as to say this is one of the best computer science books I've read. It's very readable and quite thorough.

For more information on using kgdb for kernel-level debugging, visit *http://kgdb.linsyssoft.com/*.

CHAPTER 8

Embedded Networking

"Give a man a fish and you feed him for a day. Teach him to use the Net and he won't bother you for week."

Everything is connected to the Internet, even refrigerators[1]. So it's time to turn our attention to network programming in the embedded space. Linux, as a Unix derivative, has extensive support for networking.

We'll begin by looking at the fundamental network interface, the socket. With that as a foundation, we'll go on to examine how common application-level network protocols can be used in embedded devices.

Sockets

The "socket" interface, first introduced in the Berkeley versions of Unix, forms the basis for most network programming in Unix systems. Sockets are a generalization of the Unix file access mechanism that provides an endpoint for communication either across a network or within a single computer. A socket can also be thought of as extension of the named pipe concept that explicitly supports a client/server model wherein multiple clients may be attached to a single server.

The principal difference between file descriptors and sockets is that a file descriptor is bound to a specific file or device when the application calls open() whereas sockets can be created without binding them to a specific destination. The application can choose to supply a destination address each time it uses the socket, for example when sending datagrams, or it can bind the destination to the socket to avoid repeatedly specifying the destination, for example when using TCP.

---

1 *www.exn.ca/Stories/2003/04/22/55.asp* or just Google "internet refrigerator." I don't suppose there are any net-connected toaster ovens yet.

Both the client and server may exist on the same machine. This simplifies the process of building client/server applications. You can test both ends on the same machine before distributing the application across a network. By convention, network address 127.0.0.1 is a "local loopback" device. Processes can use this address in exactly the same way they use other network addresses.

Try it out

Execute the command /sbin/ifconfig. This will list the properties and current status of network devices in your system. You should see at least two entries: one for eth0, the Ethernet interface and the other for lo, the local loopback device.

This command should work on both your development host and your target with similar results. ifconfig, with appropriate arguments, is also the command that sets network interface properties.

The Server Process

Figure 8-1 illustrates the basic steps that the server process goes through to establish communication. We start by creating a socket and then bind() it to a name or destination address. For local sockets, the name is a file system entry often in /tmp or /usr/tmp. For network sockets it is a *service identifier* consisting of a "dotted quad" Internet address (as in 192.168.1.11 for example) and a protocol port number. Clients use this name to access the service.

Next, the server creates a connection queue with the listen() service and then waits for client connection requests with the accept() service. When a connection request is received successfully, accept() returns a new socket, which is then used for this connection's data transfer. The server now transfers data using standard read() and write() calls that use the socket descriptor in the same manner as a file descriptor. When the transaction is complete the newly created socket is closed.

The server may very well spawn a new process to service the connection while it goes back and waits for additional client requests. This allows a server to serve multiple clients simultaneously. Each client request spawns a new process with its own socket.

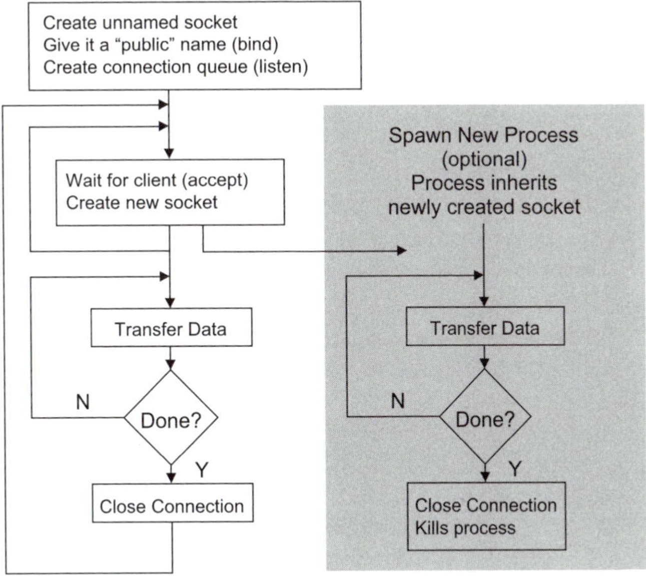

Figure 8-1: The Server Process

The Client Process

Figure 8-2 shows the client side of the transaction. The client begins by creating a socket and naming it to match the server's publicly advertised name. Next, it attempts to connect() to the server. If the connection request succeeds, the client proceeds to transfer data using read() and write() calls with the socket descriptor. When the transaction is complete, the client closes the socket.

If the server spawned a new process to serve this client, that process should go away when the client closes the connection.

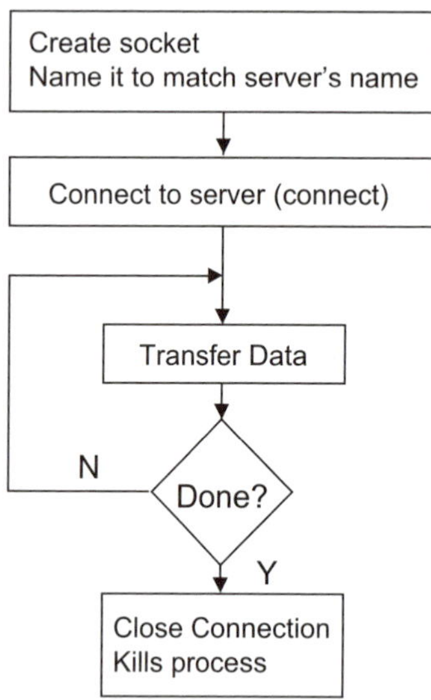

Figure 8-2: The Client Process

Socket Attributes

The socket system call creates a socket and returns a descriptor for later use in accessing the socket.

```
#include <sys/socket.h>
int socket (int domain, int type, int protocol);
```

A socket is characterized by three attributes that determine how communication takes place. The *domain* specifies the communication medium. The most commonly used domains are PF_UNIX for local file system sockets and PF_INET for Internet connections. The "PF" here stands for Protocol Family.

The domain determines the format of the socket name or address. For PF_INET, the address is AF_INET and is in the form of a dotted quad. Here "AF" stands for Address Family. Generally there is a 1 to 1 correspondence between AF_ values and PF_ values. A network computer may support many different network services. A specific service is identified by a "port number." Established network services like ftp, http, etc have defined port numbers, usually below 1024. Local services may use port numbers above 1023.

Some domains, PF_INET for example, offer alternate communication mechanisms. SOCK_STREAM is a sequenced, reliable, connection-based, two-way byte stream. This is the basis for TCP and is the default for PF_INET domain sockets. SOCK_DGRAM is a *datagram* service. It is used to send relatively small messages with no guarantee that they will be delivered or that they won't be reordered by the network. This is the basis of UDP. SOCK_RAW allows a process to access the IP protocol directly. This can be useful for implementing new protocols directly in User Space.

The protocol is usually determined by the socket domain and you don't have a choice. So the protocol argument is usually zero.

A Simple Example

The Server

cd /opt/BlueCat/demo.x86/apps/network and open the file netserve.c. First we create a server_socket that uses streams. Next we need to bind this socket to a specific network address. That requires filling in a sockaddr_in structure, server_addr. The function inet_aton() takes a string containing a network address as its first argument, converts it to a binary number and stores it in the location specified by the second argument, in this case the appropriate field of server_addr. Oddly enough, inet_aton() returns *zero* if it succeeds. In this example the network address is passed in through the compile-time symbol SERVER so that we can build the server to run either locally through the loopback device or across the network.

The port number is 16 bits and is arbitrarily set to 4242. The function htons() is one of a small family of functions that solves the problem of transferring binary data between computer architectures with different byte ordering policies. The Internet has established a standard "network byte order," which happens to be Big Endian. All binary data is expected to be in network byte order when it reaches the network. htons() translates a short (16 bit) integer from "host byte order," whatever that happens to be, to network byte order. There is a companion function, ntohs() that translates back from network byte order to host order. Then there is a corresponding pair of functions that do the same translations on long (32 bit) integers[2].

Now we *bind* server_socket to server_addr with the bind() function. Finally, we create a queue for incoming connection requests with the listen() function. A queue

[2] Try to guess the names of the long functions.

length of one should be sufficient in this case because there's only one client that will be connecting to this server.

Now we're ready to *accept* connection requests. The arguments to accept() are:

- the socket descriptor;
- a pointer to a sockaddr structure that accept() will fill in;
- a pointer to an integer that currently holds the length of the structure in argument 2. accept() will modify this if the length of the client's address structure is shorter.

accept() blocks until a connection request arrives. The return value is a socket descriptor to be used for data transfers to or from this client. In this example the server simply echoes back text strings received from the client until the incoming string begins with "q".

The Client

Now look at netclient.c. netclient determines at run time whether it is connecting to a server locally or across the network. We start by creating a socket and an address structure in the same manner as in the server. Then we *connect* to the server by calling connect(). The arguments are:

- the socket descriptor;
- a pointer to the sockaddr structure containing the address of the server we want to connect to;
- The length of the sockaddr structure.

When connect() returns we're ready to transfer data. The client prompts for a text string, writes this string to the socket and waits to read a response. The process terminates when the first character of the input string is "q".

Try it Out

To build the client and server to run on the local host, do the following:

```
make client
make server
```

Open another terminal window. In this window cd to the network/ directory and execute ./netserve. Go back to the original window and execute ./netclient. Type in a few strings and watch what happens. To terminate both processes, enter a string that begins with "q" ("quit" for example).

Both the server and client are built with debugging information on so you can run either or both of them under DDD.

Next we'll want to run netserve on the target with netclient running on the host. Execute the command:

```
make server SERVER=REMOTE
```

In the terminal window connected to the target (the one running minicom), cd to the network/ directory and execute ./netserve. Back in the original window execute ./netclient remote.

A Remote Thermostat

Moving on to a more practical example, our thermostat may very well end up in a distributed industrial environment where the current temperature must be reported to a remote monitoring station and setpoint and limit need to be remotely settable. Naturally we'll want to do that over a network. The network/ directory includes a "network-ready" version of thermostat. And if you had difficulty with the programming assignment in Chapter 6 to make the parameters settable through the serial port, this version of thermostat follows the same basic strategy.

Open network/thermostat.c. This file ends up creating, at a minimum, three processes:

1. The basic thermostat process that controls the heater and alarm.
2. A net server that accepts connections on behalf of clients that need to know the current temperature or change operational parameters.
3. A monitor process created for each network connection accepted by the server. This process parses and acts on commands sent by the client.

Start down in main() at line 195 where we get and attach a shared memory region for the thermostat's operational parameters and data. After initializing the operational parameters this process forks. The parent process goes on to run the thermostat as originally defined back in Chapter 5. The child process invokes a function called server().

Moving up to around line 115, server() is pretty much a duplication of the code we saw earlier in netserve.c. It creates and names a socket, binds to the socket, and sets up a connection queue. Then it waits to accept connections on behalf of network clients. The difference is that when server() accepts a network connection, it forks a new monitor() process to service that connection. It then goes back to wait for additional network connections.

This means that the thermostat is capable of responding to multiple clients from anywhere in the network. In a real world application you would probably impose additional restrictions such that any client could request the current temperature, but only "trusted" clients would be allowed to change the operational parameters such as setpoint and limit. This could be implemented, for example, by creating two network servers at different port numbers. One server would simply supply current temperature values. The other, responding only to trusted clients, would allow for changing the operational parameters. The implementation of this functionality is left as the proverbial "exercise for the student."

The monitor() process begins at line 57. The first thing it does is get and attach the shared memory region created by the main thermostat process. Then it just reads and acts on commands coming in on the client_socket. Note the use of the client_socket integer variable. It is a global variable in the file thermostat.c. monitor() gets the value of client_socket at the time that the server() process forks. Remember that the value of client_socket seen by monitor() is a *copy* and so from the viewpoint of monitor() it doesn't change no matter what subsequently happens in server().

The server() process goes back and waits for another client connection. When a new client request arrives, accept() will return a different value to be stored in client_socket. This is the value seen by the new monitor() process when server() forks.

monitor() adds some commands on top of what we discussed in Chapter 5:

- "d" <number> sets the deadband.
- "?"sends back the current temperature.
- "q" terminates the connection and kills the corresponding monitor process.

We might have structured monitor() to automatically send the current temperature to the client at regular intervals in the same way the thermostat currently reports temperature to stdout. But that would have necessitated a different client strategy.

The strategy of asking for the current temperature allows us to use netclient as presently constituted to access the thermostat. Nevertheless, feel free to play around with an implementation of both monitor() and netclient that automatically returns the current temperature at client-specified intervals.

The Makefile in network/ includes both sim and target targets to build the networked thermostat in both the simulation and target environments. To build the simulation version simply execute

```
make sim
```

To run the simulation version you'll need three terminal windows as follows:

1. Runs devices from the thermostat/ directory.
2. Runs thermostat.s from the network/ directory.
3. Runs netclient from the network/ directory.

thermostat is built with the –g compiler flag so you can use DDD to investigate its operation.

When you're happy with the simulation version, you can build the target version of thermostat with

```
make target SERVER=REMOTE
```

Before running thermostat.t on the target, make sure the parport device driver is loaded.

Embedded Web Servers

HTTP, the protocol of the World Wide Web, has become the de facto standard for transferring information across networks. No big surprise there. After all, virtually every desktop computer in the world has a web browser. Information rendered in HTTP is accessible to any of these computers with no additional software.

Background on HTTP

Hypertext transfer protocol (HTTP) is a fairly simple synchronous request/response ASCII protocol over TCP/IP as illustrated in Figure 8-3. The client sends a request message consisting of a header and, possibly, a body separated by a blank line. The header includes what the client wants along with some optional information about its capabilities. Each of the protocol elements shown in Figure 8-3 is a line of text terminated by CR/LF. The single blank line at the end of the header tells the server to proceed.

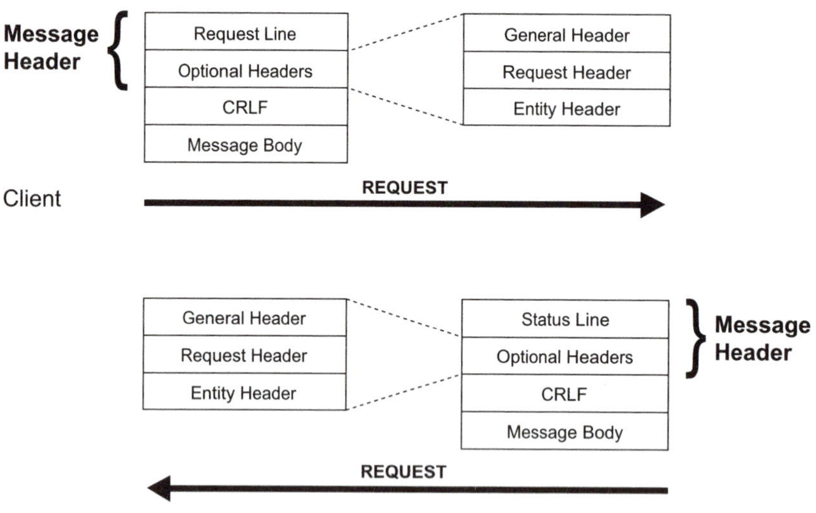

Figure 8-3: HTTP Protocol

A typical request packet is shown in Listing 8-1. The first line starts with a *method token*, in this case GET, telling the server what "method" to use in fulfilling this request. This is followed by the "resource" that the method acts on, in this case a file name. The server replaces the "/" with the default file index.html. The remainder of the first line says the client supports HTTP version 1.1

```
GET / HTTP/1.1
Host: 127.0.0.1
User-Agent: Mozilla/5.0 (X11; U; Linux i586; en-US; rv:1.2.1)
Gecko/20030225
Accept: text/xml,application/xml,application/xhtml+xml,text/
html;q=0.9,text/plain;q=0.8
Accept-Language: en-us, en;q=0.50
Accept-Encoding: gzip, deflate, compress;q=0.9
Accept-Charset: ISO-8859-1, utf-8;q=0.66, *;q=0.66
Keep-Alive: 300
Connection: keep-alive
<blank line>
```

Listing 8-1

The Host: header specifies to whom the request is directed, while User-Agent: header identifies who the request is from. Next come several headers specifying what sorts of things this client understands in terms of media types, language, encoding, and character sets. The Accept: header line is actually much longer than shown here.

The Keep-Alive: and Connection: headers are artifacts of HTTP version 1.0 and specify whether the connection is "persistent," or is closed after a single request/response interaction. In version 1.1 persistent connections are the default. This example is just a small subset of the headers and parameters available. For our fairly simple embedded server, we can in fact ignore most of them.

A Simple Embedded Web Server

To make data available via HTTP you need a web server. Creating a web server for an embedded device is not nearly as hard as you might think. That's because all the rendering, the hard part, is done by the client, the web browser. By and large, all the server has to do is deliver a file.

The network/ directory contains a simple example called webserve.c. Have a look with your favorite editor. It starts out quite similar to netserve.c except that it listens on port 80, the one assigned to HTTP. Once the connection is established, the server reads a request message from the client and acts on it. For our rather limited purposes, we're only going to handle two methods: POST and GET.

In the case of GET, the function doGETmethod() near line 188 opens the specified file and determines its content type. In "real" web servers like Apache, HTML files are kept in a specific directory. In this case it just seems easier to put the files in the same directory as the program, so doGETmethod() strips off the leading "/" if present to make the path relative to the current directory.

If everything is ok, we call responseHeader() to send the success response. The response header indicates the content type of the file and also tells the server that we want to close the connection, even if the client asked to keep the connection alive. Finally, we send the file itself. If it's an HTML file we need to parse it looking for dynamic content tags.

Dynamic Web Content

Just serving up static HTML web pages isn't particularly interesting, or even useful, in embedded applications. Usually the device needs to report some information and we may want to exercise some degree of control over it. There are a number of ways to incorporate dynamic content into HTML, but for our limited purposes, we're going to take a "quick and dirty" approach.

This is a good time to do a quick review of HTML. Take a look at the file index.html.

A nice feature of HTML is that it's easily extensible. You can invent your own tags. Of course, any tag the client browser doesn't understand it will simply ignore. So if we invent a tag, it has to be interpreted by the server before sending the file out. We'll invent a tag called <DATA> that looks like this:

```
<DATA data_function>
```

index.html has one data tag in it.

The server scans the HTML text looking for a <DATA> tag. data_function is a function that returns a string. The server replaces the <DATA> tag with the returned string. Open webvars.c with an editor. Near the top is a table with two entries each consisting of a text string and a function name. Just below that is the function cur_time(), which returns the current time as an ASCII string.

Now go back to webserve.c and find the function parseHTML() around line 129. It scans the input file for the <DATA> tag. When one is found, it writes everything up to the tag out to the socket. Then it calls web_var() with the name of the associated data function. web_var(), in webvars.c, looks up and invokes the data function, and returns its string value. The return value from web_var() is then written out and the scan continues.

Needless to say, a data function can be anything we want it to be. This particular example happens to return the current time. Incidentally, there's a subtle bug in parseHTML(). See what you can do about it[3].

Forms and the POST Method

The <DATA> tag is how we send dynamic data from the server to the client. HTML includes a simple mechanism for sending data from the client to the server. You've no doubt used it many times while web surfing. It's the <FORM> tag. The one in our sample index.html file looks like this:

---

[3] Hint: What happens if a <DATA> tag spills over a buffer boundary? An expedient way around this is to read and write one character at a time; but as we saw in Chapter 7, the low-level I/O API incurs significant overhead, so it's better if we can read and write in larger chunks.

```
<FORM ACTION="control" METHOD="POST">
    <INPUT TYPE=TEXT NAME="Control1" SIZE=10>Control 1:
    <INPUT TYPE=TEXT NAME="Control2" SIZE=10>Control 2:
    <INPUT TYPE=TEXT NAME="Control3" SIZE=10>Control 3:
    <INPUT TYPE="submit" NAME="Go">
</FORM>
```

This tells the browser to use the POST method to execute an ACTION named "control" to send three text variables when the user presses a "Go" button. Normally the ACTION is a CGI script. That's beyond the scope of this discussion so we'll just implement a function for "control."

Have a look at the doPOSTmethod() function around line 170. It retrieves the next token in the header, which is the name of the ACTION function. Then it calls web_var(), the same function we used for the dynamic <DATA> tag. In this case, we need to pass in another parameter, a pointer to the message body because that's where the data strings are.

In the <DATA> tag case, a successful return from web_var() yields a string pointer. In the POST method case, there's no need for a string pointer, but web_var() still needs to return a nonzero value to indicate success.

The body of a HTTP POST message contains the values returned by the client browser in the form:

```
<name1>=<value1>&<name2>=<value2>&<name3>=<value3>&Go=
```

The function parseVariable() in webvars.c searches the message body for a variable name and returns the corresponding value string.

Build and Try It

Go ahead and build the web server with make webserve. Run it under DDD on your workstation so you can watch what happens as a web browser accesses it. You must be running as root in order to bind to the HTTP port, 80.

In your favorite web browser enter:

http://127.0.0.1

into the destination window. You should see something like Figure 8-4 once the file index.html is fully processed.

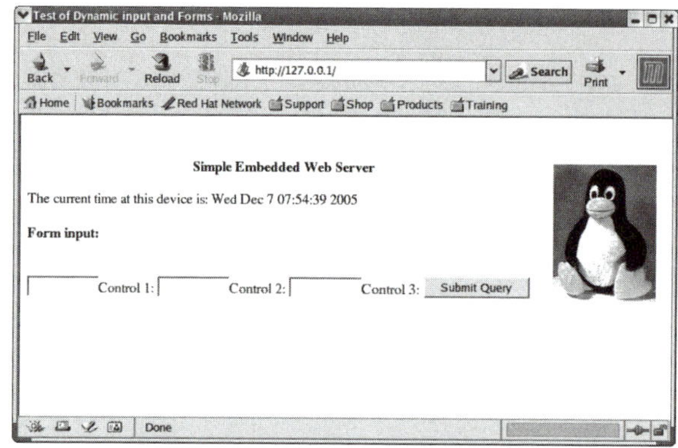

Figure 8-4: Embedded Web Server

A Web-Enabled Thermostat

Your next programming assignment is to upgrade the remote thermostat program to be a web server. All the pieces are there. You just have to put them together right. Let's try a different approach this time. It turns out that the web server requires only a trivial change to work with the thermostat. So I suggest we leave it as a separate program rather than build it into thermostat.c as we did earlier.

Here's a brief summary, by source file, of what needs to change to create a web-enabled thermostat.

- *webserve.c* – The dynamic data and forms functions will need to access the thermostat's shared memory area. At the top of main(), call a function that attaches to the shared memory. The function itself should be in webvars.c. That's the only change to webserve.c.
- You might also consider putting in the code that spawns a new process when a connection is established. This would allow simultaneous access from multiple client browsers.
- *webvars.c* – Add the function to attach to the thermostat's shared memory. Modify the dynamic data function to return current temperature. You'll probably want to change the function name. Modify the POST handler to get "Setpoint", "Limit", and "Deadband", convert them to numbers and put them in shared memory.

- *thermostat.c* – Remove all references to networking. There's nothing to fork. This reduces the file to a signal handler and `main()`, which is now just the thermostat state machine.
- *Index.html* – Change the function name in the <DATA> tag to match what you did in `webvars.c`. Likewise, change the variable names in the <FORM> tag.

Rebuild `webserve` and `thermostat`. For debugging purposes you probably want to run each of them in a separate window on the workstation. On the target you'll need to run one, or both, in the background.

Once you have that running on both the workstation and target, here's an "extra credit" assignment. There are three small GIF files in the `network/` directory that are circles—one red, one green, and one open. These can be used to represent the states of the heater and alarm. So try adding a couple of <DATA> tags that will put out the appropriate circle image based on the heater and alarm states.

Of course, the other thing we would want to do is dynamically update the temperature reading. Dynamic web content gets into java applets and other issues that are beyond the scope of this book, but are definitely worth pursuing. Google the phrase "embedded web server" to find a veritable plethora of commercial products and open source projects that expand on what we've done here.

Embedded email

To wrap up our tour of embedded networking, let's take a quick look at how we might use email in an embedded context. Suppose our thermostat maintains a log of temperatures and tracks how often the heater comes on and for how long. We might want to periodically send that log someplace for review and analysis.

The protocol for sending email is called simple mail transfer protocol (SMTP). Like HTTP it is a fairly simple ASCII-based client server mechanism. SMTP uses a full duplex stream socket where the client sends text commands and the server replies with responses containing numeric status codes. It is a "lock step" protocol in which every command must get a response.

Listing 8-2 is a typical dialog with an SMTP server. The client entries are bolded and extra blank lines have been added to improve readability. When the connection is first established, the server introduces itself telling us a little about it. The client in turn introduces itself with the HELO command, specifying a domain name. Note that each reply from the server begins with a 3-digit status code. This allows for straightforward automatic handling of a message dialog, yet the accompanying text supports human interpretation. Commands and replies are not case sensitive.

Next we specify the sender (FROM) and recipient (RCPT) of the message. There can be multiple recipients. The SMTP spec calls for email addresses to be enclosed in angle brackets, at least that's what the examples show. Yet my SMTP server seems quite happy without them. The body of the message is introduced with the DATA command. The server responds to the DATA command and then doesn't respond again until the client sends the termination sequence, which is a "." on a line by itself. At this point we could send another message or, as in this case, QUIT. The server closes the connection and so should the client.

The program maildialog lets you interact with a SMTP server. That's where Listing 8-2 came from. Have a look at maildialog.c. The program accepts one runtime argument, the name of a SMTP mail server that can be either a dotted quad or a fully qualified domain name. Note the function gethostbyname(). The argument to this function is a host identifier, either a dotted quad represented as a string, or a fully qualified domain name. In the latter case, gethostbyname() goes out to the network to find the address of the specified domain name.

After opening a socket to the server, the program drops into a loop where it:

1. Prints the previous server response.
2. Checks for a couple of key words in the previous command that alter the loop's behavior.
    - For the DATA command, drop into another loop that sends text until it sees a line that just has "."
    - For the QUIT command, break out and close the connection.
3. Gets another command from the user.

Build and try it with your own email server.

```
make maildialog
maildialog <your_server_name>
```

The file sendmail.c implements a simple client that can be called from an application to send an email. The mail server and local domain names are specified in string pointers that must be declared by the calling application. The message itself is embodied in a data structure. The application simply fills in the data structure and calls sendMail().

Note the do {…} while (0) construct. This is a way of getting out of the message transaction if something goes wrong without using the dreaded goto. In case of an error, we simply break out of the while loop.

Try it out. Create a new process for the thermostat that wakes up, say, every 15 minutes and sends an email with the current temperature.

```
220 mail.cybermesa.com ESMTP Sendmail 8.12.11/8.12.10; Tue, 13
Dec 2005 07:29:36-0700

helo mydomain.com
250 mail.cybermesa.com Hello [127.19.50.42], pleased to meet you

mail from: <me@mydomain.com>
250 2.1.0 <me@mydomain.com>... Sender ok

rcpt to: doug@intellimetrix.us
250 2.1.5 doug@intellimetrix.us... Recipient ok

data
354 Enter mail, end with "." on a line by itself

content-type: text
Subject: This is a test
this is a test of interactive email

Doug
.
250 2.0.0 jBDETaDE025477 Message accepted for delivery

quit
221 mail.cybermesa.com closing connection
```

Listing 8-2

Email isn't limited to just text. Using MIME (Multipurpose Internet Mail Extensions), we can send virtually any kind of binary data—pictures, audio, video, firmware updates—as attachments. Like most Internet protocols, SMTP requires that arbitrary binary data be translated into 7-bit ASCII before transmission. This is usually done with Base64 encoding.

The flip side of sending an email, of course, is receiving it. The most popular mechanism in use today for receiving email is called the Post Office Protocol, or POP currently at version 3, known as *POP3*. POP3 is a command/response protocol quite similar to SMTP as illustrated in Listing 8-3, which happens to show the receive side of the send transaction shown in Listing 8-2.

What role might an embedded POP3 client serve? Automating the receipt and analysis of the log sent by our networked thermostat comes to mind. This scenario isn't necessarily a classic "embedded" application, it might very well run on a workstation. Nevertheless, it could be an element in a larger embedded application.

Like SMTP, a POP3 transaction begins with introductions, which in this case include a password. There are a few commands, among which are LIST and RETR, which return multiple line responses terminating in a line with just a period. Having retrieved a message, we can delete it from the server. And, of course, when we're all done, we QUIT, closing the connection to the server.

It's a relatively trivial exercise to modify **maildialog.c** to interact with a POP3 server instead of a SMTP server. So your extra credit assignment for this section is exactly that. Create a version of **maildialog** that interacts with a POP3 server to retrieve email. POP3 servers normally listen at Port 110.

Turns out there are a couple of "gotchas" related to searching for the termination string, "." on a line by itself. First, the server may send multiple lines as a single string, so you can't rely on the period being the first character. You need to search the string for "\n." Second, to verify that the period is on the line by itself, it's not sufficient to just search for "\n.\n" because the server may insert a carriage return, "\r", character. So following the "\n." you need to look for either "\n' or "\r".

Other Application-Level Protocols

The three protocols we've looked at, HTTP, SMTP, and POP3, can be useful for certain classes of embedded problems. There are a great many other application-level protocols that can also be useful in other situations. SNMP (Simple Network Management Protocol), for example, has become the de facto standard mechanism for managing network resources. Virtually every device attached to a network implements SNMP and your embedded devices probably should too. In an embedded scenario, SNMP can be used for simple monitoring and control.

```
+OK X1 NT-POP3 Server dpmail12.doteasy.com (IMail 8.05 2264142-4)

user doug%intellimetrix.us
+OK send your password

pass *****
+OK maildrop locked and ready

list
+OK 1 messages (647 octets)

1 647
.

retr 1
+OK 647 octets

Received: from mail.cybermesa.com [198.59.109.2] by dpmail12.doteasy.
com with ESMTP
  (SMTPD32-8.05) id AAEA563800F4; Tue, 13 Dec 2005 06:30:02 -0800
Received: from mydomain.com ([127.19.50.42])
        by mail.cybermesa.com (8.12.11/8.12.10) with SMTP id jB-
DETaDE025477
        for doug@intellimetrix.us; Tue, 13 Dec 2005 07:31:09 -0700
Date: Tue, 13 Dec 2005 07:29:36 -0700
From: me@mydomain.com
Message-Id: <200512131431.jBDETaDE025477@mail.cybermesa.com>
content-type: text
Subject: This is a test
X-IMAIL-SPAM-VA
LFROM: (1446510836)
X-RCPT-TO: <doug@intellimetrix.us>
Status: RX-UIDL: 329549330

This is a test of interactive email.

Doug
.

dele 1
+OK msg deleted

quit
+OK POP3 Server saying Good-Bye
```

Listing 8-3

Resources

Linux Network Administrators' Guide, available from the Linux Documentation Project, *www.tldp.org*. Not just for administrators, this is a quite complete and quite readable tutorial on a wide range of networking issues.

Jones, M. Tim, *TCP/IP Application Layer Protocols for Embedded Systems*, Charles River Media, 2002. The idea of adding a <DATA> tag to HTML came from this book. It covers a wide range of application-level network protocols that can be useful in embedded scenarios.

Comer, Douglas, *Internetworking with TCP/IP, Vols. 1, 2 and 3*, Prentice-Hall. This is the classic reference on TCP/IP. Volume 1 is currently up to its fifth edition dated 2005. Volume 2, coauthored with David L. Stevens, is at the third edition, 1998, and volume 3, also coauthored with Stevens, dates from 1997. Highly recommended if you want to understand the inner workings of the Internet.

Internet protocols are embodied in a set of documents known as *RFCs*, *Request for Comment*. The RFCs are now maintained by a small group called the RFC Editor. The entire collection of RFCs, spanning the 30-plus year history of the Internet, is available from *www.rfc-editor.org*. In particular, HTTP is described in RFC 2616, SMTP is described in RFC 821, and POP3 is described in RFC 1081.

CHAPTER 9

Introduction to Real-Time Programming

"But there never seems to be enough time to do the things you want to do once you find them."
— Jim Croce, Time in a Bottle

Fundamentally, "real-time" means that a program must respond to *events* in its environment within a specified deadline. Such systems are said to be *event-driven* and can be characterized in terms of *latency* where latency is defined as the time interval from when an event occurs until the time the system takes action in response to that event. In a general-purpose operating system such as Windows or Unix, latency is of little concern. As a user of Windows, you probably couldn't tell the difference if the system responds to a key press in 20 milliseconds or 220 milliseconds. And if the system happens to be doing something else, it may take two seconds to respond. You may or may not notice, but it happens rarely enough that you probably won't be too upset.

Real-time, on the other hand, demands an upper limit on latency, also called the *scheduling deadline*. Real-time systems can be roughly divided into two major classes: *hard* real-time and *soft* real-time. The distinction is that in hard real-time, the system absolutely must meet its scheduling deadline each and every time. Failure to meet the deadline may have catastrophic consequences including loss of life. A fly-by-wire aircraft control system is an example of hard real-time. The control algorithms depend on regular sampling intervals. If sampling is delayed, the algorithm could become unstable.

Consider the aforementioned fly-by-wire system. Suppose the system senses that the plane is losing altitude. It responds by increasing power to the engines, which will reduce the rate of decent. The control algorithm is programmed with a specific relationship between power level and rate of climb or descent for that specific aircraft.

Now suppose that the next sample is delayed, maybe because Windows is busy putting up the paper clip icon. The next sample will report a higher altitude than what would have been reported if the sample had been taken at the correct time. So the algorithm erroneously reduces engine power to compensate. At the next sample time the reported altitude will be lower, but may be too low if again the sample time was delayed.

In soft real-time, the scheduling deadline is more of a goal than an absolute requirement. We expect the system to meet its deadline most of the time, but nothing particularly bad happens if it's occasionally late. Failure to meet the deadline simply results in degraded performance without catastrophic consequences. The automated teller network is a good example of soft real-time. Is the ATM network real-time? You bet it is! When you put your ATM card into the machine, you expect a response within a couple of seconds. But if it should take longer, the worst that happens is you get impatient.

Many systems exhibit both kinds of behavior. That is, some parts are hard and some parts are soft.

Polling vs. Interrupts

Real-time systems are said to be "event-driven," meaning that a primary function of the system is to respond to "events" that occur in the system's environment. How does the program respond to events? There are two fundamental approaches. The first is *polling* as illustrated in Listing 9-1. The program begins with some initialization and then enters an infinite loop that tests each possible event to which the system must respond. For each event that is set, the program invokes the appropriate servicing function.

```
int main (void)
{
sys_init();
while (TRUE)
{
        if (event_1)
            service_event_1();
        if (event_2)
            service_event_2();
```

Listing 9-1: The Polling Loop *(continued)*

```
            .
            .
        if (event_n)
            service_event_n();
    }
}
```

Listing 9-1: The Polling Loop

This strategy is simple to implement and quite adequate for small systems with relatively loose response time requirements. But there are some obvious problems:

- The response time to an event varies widely depending on where in the loop the program is when the event occurs. For example, if event_1 occurs immediately before the if (event_1) statement is executed, the response time is very short. However if it occurs immediately *after* the test, the program must go through the entire loop before servicing event_1.
- As a corollary, response time is also a function of how many events happen to be set at the same time and consequently get serviced in the same pass through the loop.
- All events are treated as having equal priority.
- As new features, hence new events, are added to the system, the loop gets longer and so does the response time.

The second approach, making use of *interrupts*, is much more efficient and, perhaps not surprisingly, more difficult to program. The idea of the interrupt is that the occurrence of an event "interrupts" the current flow of instruction execution and invokes another stream of instructions that services the event as illustrated in Figure 9-1. When servicing is complete, control returns to where the original instruction stream was interrupted. Servicing the event happens "right now" and doesn't have to wait for the main program to "get around to it." The instruction stream that services the event is called an *interrupt service routine* or ISR.

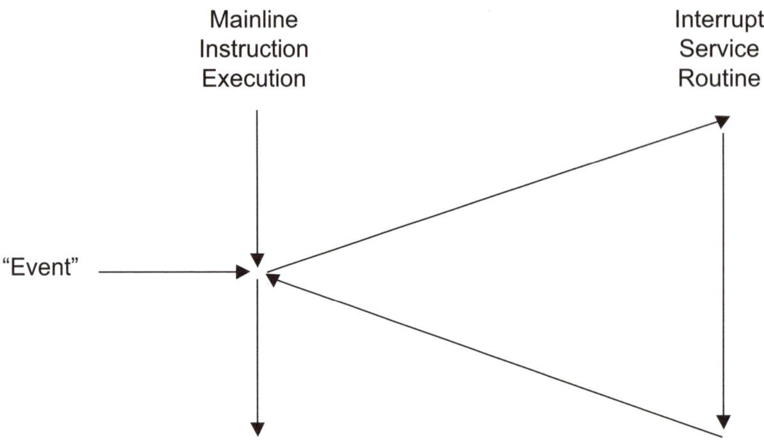

Figure 9-1: The Interrupt

Most modern processors implement three distinct types of interrupt. The first two are:

- The INT instruction, sometimes called a *TRAP*. This is like a subroutine call with one important difference as we'll see shortly.
- Processor exceptions. Fault conditions such as divide by 0 or illegal memory reference can be handled through the interrupt mechanism.

These two forms of interrupt are *synchronous* with respect to instruction execution. That is, INT *is* an instruction and fault exceptions are the direct result of instruction execution.

The third type of interrupt is generated by events that occur externally to the processor. These are generated by the I/O hardware and occur *asynchronously* with respect to instruction execution.

It is the asynchronous external interrupt that:

- Maximizes the performance and throughput of computing systems.
- Creates the most problems and frustrations for programmers[1].

Most processors utilize a similar interrupt scheme. Figure 9-2 shows how the Intel x86 architecture does it. The first 1 Kbytes of memory[2] are reserved for an *Interrupt Vector Table*. Each vector is 4 bytes representing the FAR address (segment:offset) of an ISR.

[1] Many years ago someone wrote in *Computer Magazine* of the IEEE Computer Society that, "The invention of the interrupt was perhaps the greatest disaster in the history of computer science."

[2] This explanation is based on "real mode" memory, just to keep it simple.

Some of these vectors have specific meanings defined by the processor architecture. For example, vector 0 is the divide-by-0 exception, vector 3 is a breakpoint (a single-byte INT instruction), vector 13 is the infamous General Protection Fault, and so on.

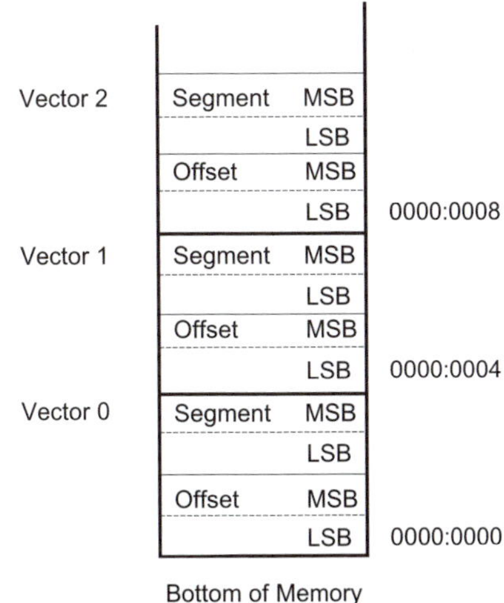

Figure 9-2: Interrupts—The Vector Table

Some vectors are reserved for external hardware interrupts. In the PC, vectors 8 to 15 and 0x70 to 0x77 are reserved for hardware.

All other vectors are accessible via 2-byte INT instructions where the second byte is the vector (or interrupt) number. System software usually establishes conventions concerning many of these vectors. For example the PC BIOS uses several interrupts for hardware services and Linux uses INT 0x80 to invoke kernel services.

Figure 9-3 illustrates the basic process of interrupt execution using the Linux INT 0x80 as an example:

- The processor saves the current Program Counter (PC) and Code Segment (CS) on the stack along with the Processor Status Word (PSW).
- The second byte of the INT instruction is an index into the vector table to find the address of the corresponding ISR. The processor loads this address into the PC and CS registers and execution proceeds from this point.
- The end of the ISR is indicated by an IRET instruction (Interrupt Return). This pops the PC, CS, and PSW off the stack so that execution resumes at the instruction following the INT.

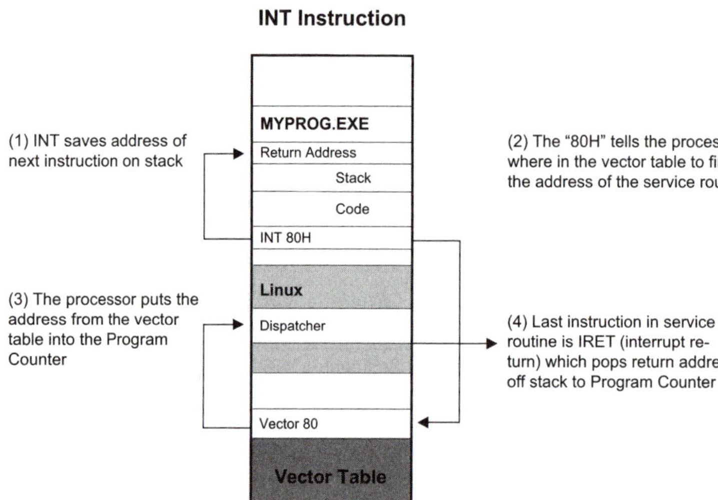

Figure 9-3: Interrupts—How They Work

The INT instruction is like the subroutine CALL instruction but with one important difference: Whereas the destination address is embedded in the CALL instruction, with the INT instruction the calling program need not know the address of the ISR! The address is held in the Interrupt Vector Table. Thus it is ideal for communication between two separately compiled and loaded programs as, for example, an application program and an operating system.

External hardware interrupts operate in a similar fashion as shown in Figure 9-4. A device requiring service asserts an *Interrupt Request (IRQ)* line. When the processor responds with *Interrupt Acknowledge (IAK)*, the device places it's interrupt vector number on the data bus. The processor then effectively simulates an INT instruction using the supplied vector index.

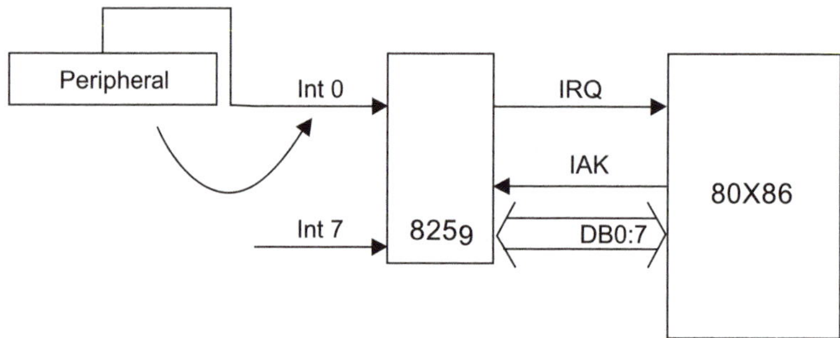

Figure 9-4: Hardware Interrupts

In practice, most systems incorporate a specialized peripheral called an *Interrupt Controller* to manage details like putting the vector index on the data bus at the correct time. The PC architecture includes two 8259 Interrupt Controllers, each able to handle up to eight interrupt inputs.

The 8259 provides a mechanism to prioritize interrupts so that more important or critical devices have precedence over less important devices.

Interrupts may also be enabled or disabled. At the processor level, interrupts may be globally enabled or disabled via the STI and CLI instructions. Individual interrupts may be selectively enabled and disabled either at the 8259 Interrupt Controller or at the device itself. In fact the ability to enable and disable interrupts is crucial to the design and implementation of real-time software.

Not surprisingly, asynchronous interrupts are not without their problems. Consider a data acquisition application based on a multichannel A/D converter as shown in Figure 9-5. Each time the A/D converter takes a set of readings on all channels, it interrupts the processor. The ISR reads the data and stores it in a memory buffer where it is available to the background main program.

Interrupt-driven operation allows us to respond to the A/D quickly while the memory buffer effectively "decouples" the background program from the data source, i.e., the background program doesn't need to know where the data comes from.

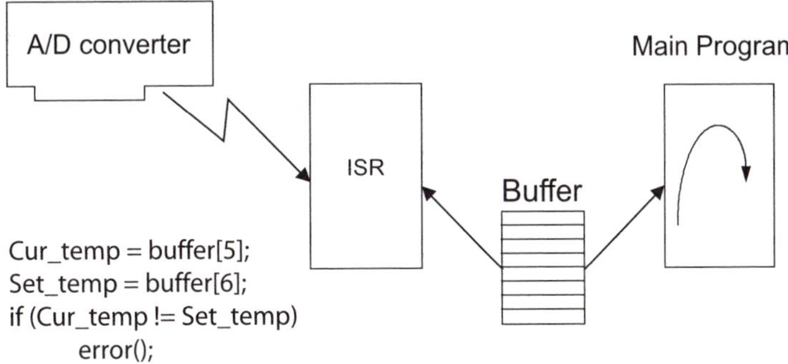

Figure 9-5: Using Interrupts—Example

Consider the code fragment shown here. Let us say that for testing purposes, we feed the same continuously varying signal into both channels 5 and 6. We would assume that the program would never fail since both channels are measuring the identical signal.

In fact, the program as written is guaranteed to fail because an interrupt can occur between the update of Cur_temp and the update of Set_temp with the result that the value for Cur_temp comes from the *previous* data set while the value for Set_temp comes from the *current* data set. Since the input signal is varying over time and the two data sets are separated by a finite time, the values will be different and the program will fail.

This then is the essence of the real-time programming problem; managing asynchronous interrupts so they don't occur at inopportune times.

There is a simple albeit inelegant solution to the problem. We can put a Disable Interrupt (CLI) instruction before the update of Cur_temp and an Enable Interrupt (STI) instruction after the update of Set_temp. This prevents the interrupt from interfering with the variable updates. It turns out that judicious use of STI and CLI is a key element of the "correct" solution but simply scattering STI and CLI throughout the code is like using go to's or global variables. It's just asking for trouble.

Tasks

The "correct" solution is called *multitasking*, which has proven to be a powerful paradigm for structuring real-time, interrupt-driven systems. I would go so far as to suggest that multitasking is first and foremost a paradigm for safely and reliably handling asynchronous interrupts. The basic idea is that we can break a large problem down into a bunch of smaller, simpler problems. Each one of these sub-problems becomes a task. Each task does one thing to keep it simple. Then we pretend that all of these tasks are running in parallel. They aren't really running in parallel unless you have a multiprocessor system. On a single processor the tasks share the processor.

Like any program, a task contains code that carries out the function the task is designed to accomplish. This code is embodied in a function that is analogous to the "main" function in a normal C program. What sets a task apart from an ordinary function is that each task has a *context* embodied in its own *stack* (See Figure 9-6).

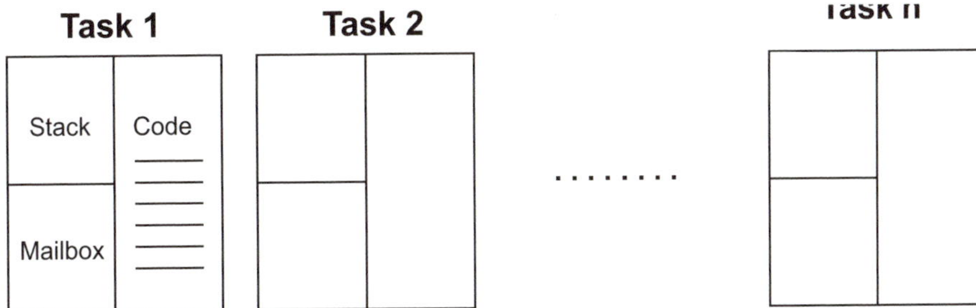

Each task consists of:

- Code to carry out the task's functionality.
- A stack to hold the task's "context".
- An optional mailbox so the task can communicate with other tasks

Figure 9-6: What Is a Task?

Note by the way that it is possible, and sometimes quite useful, to create multiple tasks from the same function. What keeps these tasks separate and distinct is that each one has its own stack. This is really classic object-oriented programming. One could think of the task function as defining a class. Then each task created from that function is an instance of the class.

Although tasks may be considered to be independent, they typically need to cooperate in order to carry out the overall mission that the system is designed for. Thus a task requires some form of communication mechanism through which it can communicate and synchronize with other tasks. For the moment, we'll call that mechanism a *mailbox*.

Listing 9-2 shows pseudo-code of a typical task. The data argument provides a way to parameterize the task in the same manner as *argc* and *argv* in main. This can be especially useful if multiple tasks are derived from the same function. The "uniqueness" of each task is conveyed by the argument value.

A task may start with some initialization (perhaps involving the data argument) after which it usually enters an infinite loop. At some point in the loop, usually near the top, it waits for "something to happen," perhaps the arrival of a message at its mailbox or simply the expiration of some time interval. While it's waiting, the task is not executing, not using the processor. Some other task that's ready to execute is using the processor.

Eventually, the event that the task is waiting for occurs. The task then "wakes up" and, if it received a message for example, decodes the message and acts on it, often with a large `switch` statement. After acting on the message, the task returns to wait for something else. Windows programmers will recognize this as the basic Windows programming model.

Note that the reason multitasking works at all is that most tasks spend most of their time waiting for something to happen.

```
void task (void *data)
{
    init_task();

    while (TRUE)
    {
        Wait for message at task mailbox();
        switch (message.type)
        {
            case MESSAGE_TYPE_X:
                ...
                break;
            case MESSAGE_TYPE_Y:
                ...
                break;
        }
    }
}
```

Listing 9-2: Typical task code

Scheduling

Tasks operate under the supervision of the real-time *kernel* that consists of:

- A collection of *services* that implement such things as inter-task communication and synchronization.
- A *scheduler* whose job it is to make sure that the highest priority ready task is the one that's currently executing.

The scheduler treats each task as a state machine. While every kernel has its own, often more complex, state model, Figure 9-7 shows conceptually the minimum state diagram for a task. The states are as follows:

- *Running*: Only one task, the currently executing task, can be in the Running state. A task can voluntarily transition from Running to Blocked by waiting on an event. In a preemptive system (we're coming to that), the scheduler may cause the Running task to transition to Ready if a higher priority task becomes Ready. This is called *preemption*.
- *Ready*: The task is ready to run but has a lower priority than the currently executing task. The task will transition from Ready to Running when it becomes the highest priority Ready task.
- *Blocked*: A Blocked task is waiting for some event to occur; a message at a mailbox, a timeout, etc. When the event occurs, the task transitions to Ready.

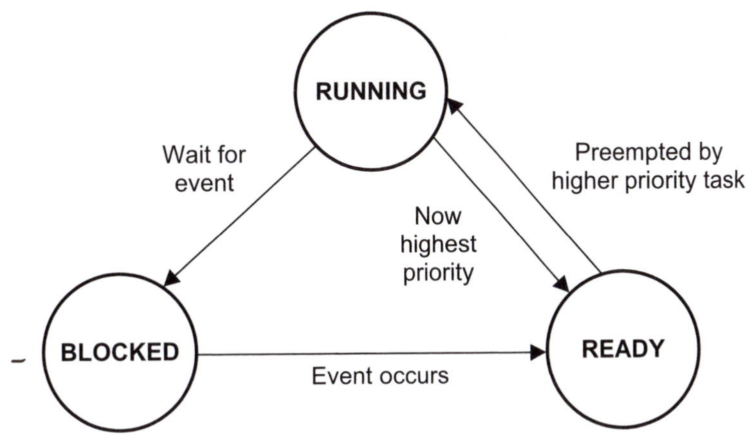

Figure 9-7: Task States

Periodic Scheduling

There are many tasks that simply need to wake up periodically, do something, and go back to sleep. There are a couple of approaches to scheduling periodic tasks as shown in Figure 9-8. Every operating system has a function called Delay(), or some variation thereof, that causes the calling task to be blocked for a specified amount of time, usually expressed in clock ticks. The top half of Figure 9-8 shows what happens when we use Delay() to schedule a periodic task, in this case with a period of three clock ticks. The behavior of the system depends on the execution time of the task. If the execution time is less than one clock tick, then the task wakes up every three ticks as desired. However if the execution time is longer than a clock tick, when the task calls Delay() it will still be blocked for three ticks. So in this example the task actually wakes up every fourth tick. That's not what we intended.

An alternative approach, not available with all systems, is to declare the task to be *periodic*. In this case the scheduler wakes the task up at the proper interval regardless of the task's execution time as shown in the bottom half of Figure 9-8. Instead of calling Delay(), a periodic task calls a function like WaitTilNext() that blocks the task until its next scheduled execution.

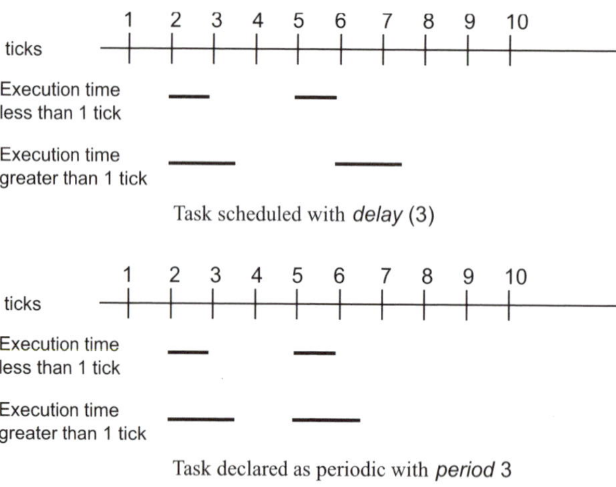

Figure 9-8: Periodic Tasks

Aperiodic Scheduling

Other tasks must respond to events occurring at random times. An event may be the arrival of a network packet, the closure of a switch that indicates a tank is full, or perhaps a conversion is complete on an analog to digital converter and it needs to be read. Very often these asynchronous events are communicated to the computer via interrupts. The ISR must have some way to communicate the occurrence of the interrupt to a task that is responsible for servicing the event. We'll see an example of that in the section on inter-task communication.

Preemptive vs. Nonpreemptive scheduling

There are two fundamental strategies for task scheduling: *preemptive* and *nonpreemptive*. Consider two tasks with the lower priority task currently Running and the higher priority task Blocked waiting for an event that will be signaled by an interrupt.

The upper half of Figure 9-9 shows what happens in the nonpreemptive case. The ISR causes the higher priority task to transition to the Ready state but at the end of the ISR, control returns to the lower priority task where it was interrupted. Later when the lower priority task blocks waiting for an event, the higher priority task becomes the Running task.

The lower half of Figure 9-9 shows the preemptive case. The difference here is that the scheduler is invoked at the end of the ISR. It determines that the higher priority task is Ready and switches tasks accordingly. The lower priority task is thus *preempted*.

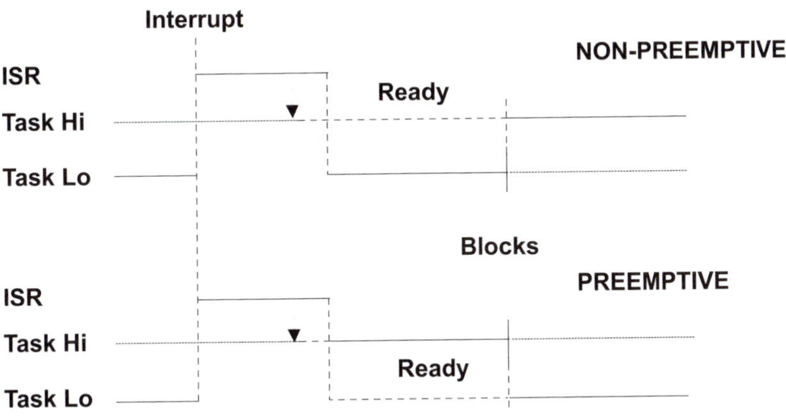

Figure 9-9: Scheduling: Nonpreemptive vs. Preemptive

A nonpreemptive system depends on all tasks being "good citizens" by voluntarily giving up the processor to be sure all tasks get a chance. Early versions of Windows were nonpreemptive. Linux is preemptive although standard Linux is not considered real-time due to excessively long periods during which preemption is disabled.

Preemptive systems provide for more predictable response times because a high priority event is serviced immediately. This is the essence of real-time—being able to guarantee the maximum time it takes to respond to an event. In the nonpreemptive case there is no guarantee how long it will be before the currently running task gives up the processor. On the other hand, preemptive systems are subject to resource conflict problems that must be carefully considered. We'll see tools for dealing with those problems shortly.

Two other scheduling strategies are employed among tasks of equal priority. In *round robin* scheduling, a task runs until it either blocks waiting for an event or voluntarily yields the processor. The distinction between blocking and yielding is that in the latter case the task is still Ready.

Consider that the Ready List contains three tasks, A, B, and C, of equal priority in that order. Task A, at the head of the list, is the Running task. When task A yields, task B becomes the Running task and the Ready List looks like this:

B C A

When task B yields, task C becomes the Running task and the list looks like this:

C A B

Thus all tasks "get a turn" provided that they all yield. Ready tasks of lower priority do not get to execute until all tasks at this level block.

Timeslicing is a variation on Round Robin that assigns a maximum time quantum or "slice" to each task to prevent one task from hogging the processor. A task runs until it blocks, yields voluntarily, or its timeslice expires. Depending on implementation, the timeslice may be the same for all processors or each task may get its own timeslice value.

Fundamentally, Round Robin scheduling is just another form of polling.

Kernel Services

A multitasking kernel is largely defined by the *services* it provides. This set of services constitutes an API that allows a user to utilize the kernel's features. When describing the functionality of a multitasking kernel, it is useful to present an API to show how the concepts map into real code. For the remainder of this chapter, I'll present a simplified idealized API that expresses the basic functionality. Real implementations will differ in detail from the model presented here but will nevertheless implement the same functionality.

Task API

Let's begin our exploration of the kernel API by examining the services required to manage tasks.

```
task_t *TaskCreate (void (*task)(void *data), void *data, int prior);
status_t TaskStart (task_t *task);
status_t TaskSuspend (task_t *task);
status_t TaskResume (task_t *task);
status_t TaskDelete (task_t *task);
```

Absolutely nothing happens in a multitasking system until we create one or more tasks. To create a task, we call a function with a name such as TaskCreate. At minimum, we have to give TaskCreate:

- a pointer to the function that implements the task's code (task);
- a pointer to the data that will be passed as the function's argument when it is first called (data);
- the task's priority (prior).

The task create service may return a pointer to a *task control block (TCB)*, identified here as type task_t. This is a data structure containing everything the kernel needs to know about the task. This pointer can then be used as an argument to other task management services.

Note that in this implementation we are assuming that the task create service allocates the TCB *and the stack*. We also assume that the stack size is fixed and is initialized somewhere else. Some implementations require the user to allocate the TCB and/or the stack and stack size. There may be additional arguments like a task name in ASCII or a timeslice value, for example.

The task create service may or may not start the task executing. If it doesn't, a separate task start service is provided. Once a task is executing it may be *suspended* which simply prevents the task from being scheduled for execution until it is subsequently *resumed*.. Finally a task that is no longer needed may be *deleted*, which removes it from the list of active tasks. In general, task management functions other than TaskCreate return a status_t type indicating whether or not the function succeeded.

Timing API

```
void Delay (unsigned int ticks);
void DelayUntil (time_t *time);
void WaitTilNext (void);
```

Every kernel has a function called something like Delay that blocks the calling task for the specified number of clock ticks. Some systems have a variation on this called DelayUntil that blocks the calling task until a specific time of day. The data type time_t is an unsigned long int representing seconds since midnight, Jan. 1, 1970.

WaitTilNext is only found in systems that support the notion of a periodic task. This function blocks the calling task until the next time the task is scheduled for execution.

Inter-Task Communication

Although tasks are considered to be independent, the overall function of the system usually requires that tasks cooperate and communicate with each other. Thus a key element of any real-time operating system is a set of communication and synchronization services.

There are several communication and synchronization mechanisms in common use:

- *Semaphore*: Used for synchronization and resource locking.
- *Event Flag*: Shows that one or more events have occurred. This is an extension of the semaphore that permits synchronizing on a combination of events.
- *Mailbox, queue, or pipe*: Mechanisms for transferring data between tasks

There are many other less widely used mechanisms such as the ADA "rendezvous" and the "monitor" in Java.

Semaphores

Consider two tasks, each of which wants to print the message "I am Task n" on a single shared printer as shown in Figure 9-10. In the absence of any kind of synchronizing mechanism, the result could be something such as "II a amm TaTasskk 12."

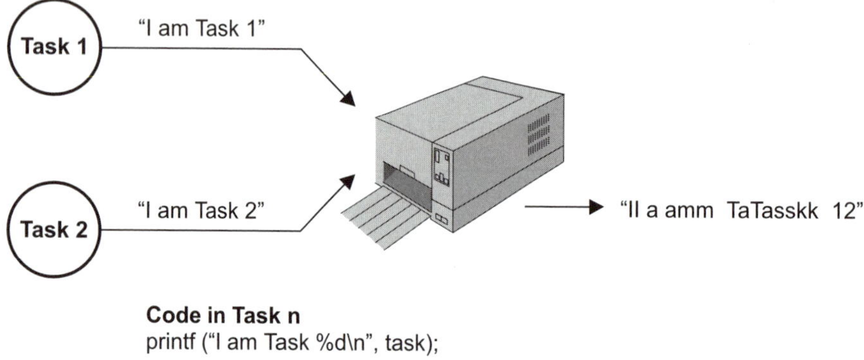

Figure 9-10: Sharing Resources

What is needed is some way to regulate access to the printer so that only one task can use it at a time.

A *semaphore* acts like a key to control access to a resource. Only the task that has the key can use the resource. In order to use the resource (in this case a printer) a task must first *acquire* the key (semaphore) by calling an appropriate kernel service (Figure 9-11). If the key is available, that is the resource (printer) is not currently in use by someone else, the task is allowed to proceed. Following its use of the printer, the task releases the semaphore so another task may use it.

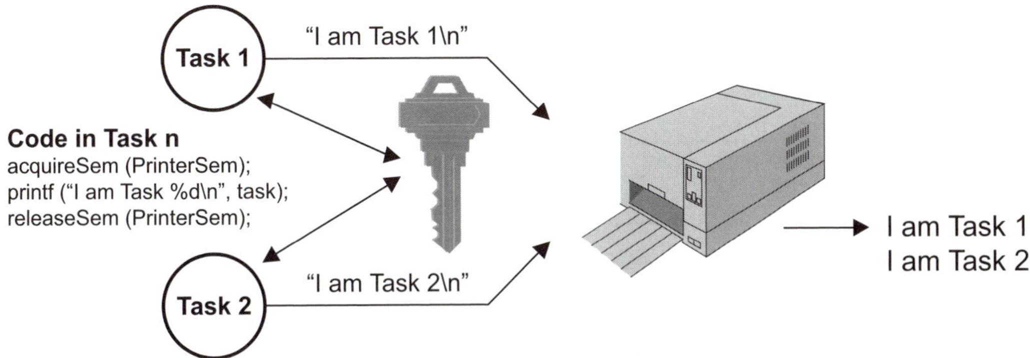

Figure 9-11: Sharing Resources with a Semaphore

If however, the printer is in use, the task is blocked until the task that currently has the semaphore releases it. Any number of tasks may try to acquire the semaphore while it is in use. All of them will be blocked. The waiting tasks are queued either in order of priority or in the order in which they called `acquireSem`. The choice of how tasks are queued at the semaphore may be built into the kernel or it may be a configuration option when the semaphore is created.

`acquireSem` works as follows:

1. Decrement the semaphore value.
2. If the resulting value is greater than or equal to 0, the resource is available and the task can proceed. Otherwise, block the task until another task executes `releaseSem`.

`releaseSem` increments the semaphore value. If the resulting value is less than or equal to 0, there is at least one task waiting for the semaphore, so make one of those tasks ready.

In the case of the printer, the semaphore was initialized to 1 reflecting the fact that there is one printer to manage. This is sometimes called a **binary** semaphore to distinguish it from the more general case of the **counting** semaphore, which can be initialized to any nonnegative number.

Consider a dynamic memory allocator that manages a fixed number of buffers as shown in Figure 9-12. Here we initialize the semaphore to the number of buffers that are initially available for allocation. When **bufReq** is called it first acquires the semaphore, then allocates a buffer. The first 10 times **bufReq** is called, the semaphore is nonnegative and the calling task proceeds. The 11th time, the calling task is blocked until someone else calls **bufRel**, which releases the semaphore.

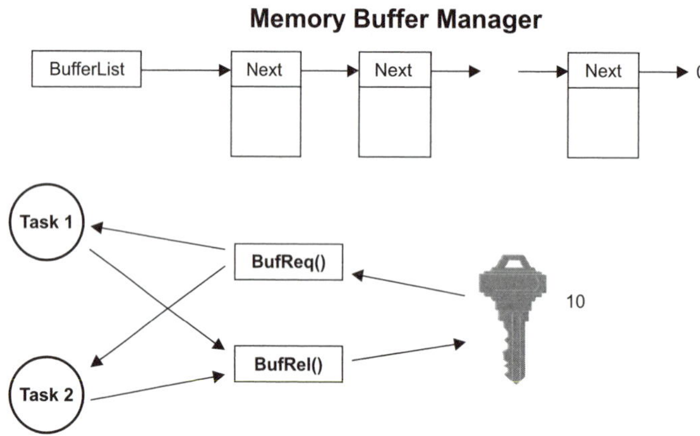

Figure 9-12: Sharing Multiple Resources

Some kernels implement both binary and counting semaphores because in some instances a binary implementation can be more efficient. The binary semaphore is sometimes called a *mutex* meaning "mutual exclusion."

A semaphore can also be used to signal the occurrence of an event as shown in Figure 9-13. For example, how does the system know that an interrupt has occurred? A task that needs to know about the occurrence of an interrupt *pends* on a semaphore. The ISR services the interrupt and then *posts* to the semaphore. (Note that the terms "pend" and "post" are used more commonly than the terms "acquire" and "release").

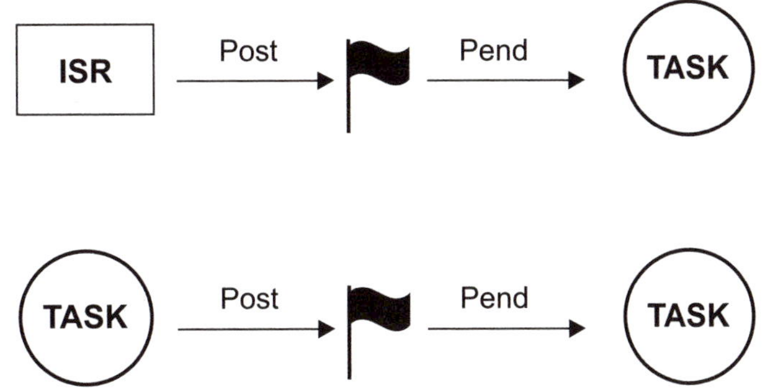

Figure 9-13: Signaling Events Through Semaphores

In the previous examples, the semaphore was initialized to a nonzero value because the resource is initially available. Here it is initialized to 0 so that when the task first pends, it is immediately blocked—the event hasn't occurred yet. When the ISR posts to the semaphore, the task "wakes up" and continues.

When a semaphore serves as a resource lock, many tasks may pend or post to it. However in the case of signaling or synchronization, the semaphore is typically used exclusively by one ISR and one task.

The same mechanism may be used by one task to signal an event to another task.

Semaphore API

```
semaphore_t *SemCreate (unsigned int value);
status_t SemDelete (semaphore_t *sem);
status_t SemPost (semaphore_t *semaphore);
status_t SemPend (semaphore_t *semaphore, unsigned int timeout);
```

A semaphore object must be created by a service typically called SemCreate that returns a pointer to the data structure, semaphore_t, that represents the semaphore. When creating a semaphore we can give it an initial value that reflects the role that this semaphore is to perform. If the semaphore is intended to protect one or more resources, the initial value is the number of resources that the semaphore is protecting. If the semaphore is intended to signal an event, it is usually given an initial value of zero because the event hasn't yet occurred. A semaphore may be deleted if it is no longer required.

The basic functions of a semaphore are SemPost and SemPend. When posting to a semaphore, the only argument typically required is the identity of the semaphore itself as returned by SemCreate. Pending on a semaphore typically requires two arguments; the identity of the semaphore and an optional timeout argument expressed in system clock ticks. If the timeout argument is nonzero, this represents the maximum time that the task is willing to wait for the semaphore to be posted. If the timeout interval expires SemPend returns an error code.

Mailboxes

In addition to signaling events, tasks often need to share data. This is usually accomplished through the mechanism of a *mailbox*. One task (the sender) posts a *message* to the mailbox while another task (the receiver) pends on the mailbox waiting for a message (Figure 9-14). If no message is posted to the mailbox when the receiving task pends on it, the task is blocked until a message is posted.

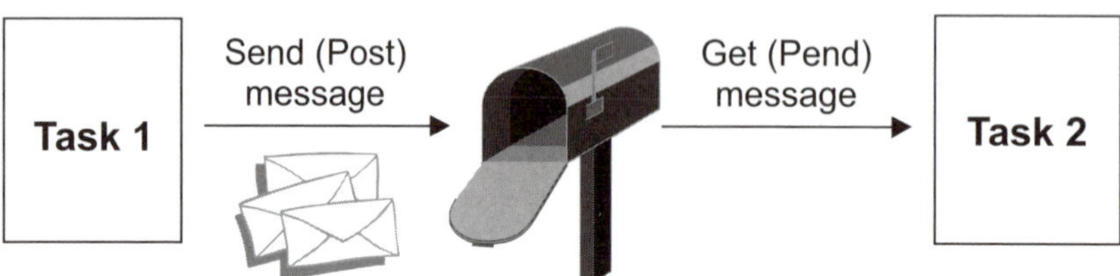

Figure 9-14: Mailboxes and Messages

In many cases a message is a pointer. What it points to must be mutually agreed upon by the two tasks involved in the transfer just as the sender and receiver of a letter must agree on what language to use. Some kernels impose a minimal amount of system structure on messages.

Of course, what the pointer points to must be accessible to both the sender and the receiver. In protected mode systems such as Linux, this is, by design, not possible. In this case the message is copied from the sender's space to the mailbox and then from the mailbox to the receiver's space.

Depending on implementation, a mailbox may hold only one message or it may be capable of queuing multiple messages that are then delivered to receiving tasks in the order in which they were sent. Some kernels offer the option of sending a "high priority" message that is immediately put at the head of the queue. As with the semaphore, any number of tasks may be waiting at the mailbox queued either in FIFO or priority order.

What happens if a task attempts to send a message to a mailbox that can only hold one message and currently contains a message? There are two implementation-dependent possibilities: the mailbox post service can return an error, or the task can be blocked until it is able to post the message. There's also the possibility that when a task posts a message to a mailbox where no one is waiting, the sender is blocked until another task receives the message.

Mailbox API

```
mailbox_t *MbxCreate (void *message);
status_t MbxDelete (mailbox_t *mbx);
status_t MbxPost (mailbox_t *mbx, void *msg);
void *MbxPend (mailbox_t *mbx, unsigned int timeout, status_t *status);
status_t MbxBroadcast (mailbox_t *mbx, void *msg);
```

The mailbox API is very similar to the semaphore API. Before we do anything else we have to create a mailbox. The mailbox create service, `MbxCreate`, returns a pointer to a structure of type `mailbox_t`. `MbxCreate` may also offer the option of sending an initial message to the mailbox when it is created.

Like the semaphore, the basic operations on a mailbox are pend and post. The arguments to `MbxPost` are the identity of the mailbox and a pointer to the message. Rather than returning status like most functions, `MbxPend` generally returns the pointer to the received message and the status variable is passed as an argument.

Some systems include a function such as `MbxBroadcast` that posts the message to all tasks that are currently pending on the mailbox.

Queues and Pipes

```
queue_t *QCreate (int qsize);
status_t QDelete (queue_t *queue);
status_t QPost (queue_t *queue, void *message);
status_t QPostFront (queue_t *queue, void *message);
void *QPend (queue_t *queue, unsigned int timeout,  status_t *status);
status_t QFlush (queue_t *queue);
```

```
pipe_t PipeCreate (void);
status_t PipeDelete (pipe_t *pipe);
status_t PipeWrite (pipe_t *pipe, void *buffer, size_t len);
status_t PipeRead (pipe_t *pipe, void *buffer, size_t len);
```

In most cases a queue is simply a mailbox that can hold multiple messages. When creating a queue we usually have to specify its size,that is the number of messages it can hold. In addition to the normal post and pend calls, a queue will usually have services to post a high priority message to the front of the queue, QPostFront (), and to flush all messages currently in the queue, QFlush ().

A pipe is a little different. Whereas mailboxes and queues move data in discrete chunks called *messages*, a pipe is generally a continuous byte stream connecting two tasks. One task reads the pipe, the other writes it. In practice, Unix systems, including Linux, treat pipes as ordinary files using the standard read() and write() functions. The pipe_t is just an array of two long integers where element zero represents the read end of the pipe and element one is the write end. The pipe is created by calling pipe ().

Problems with Solving the Resource Sharing Problem—Priority Inversion

Using semaphores to resolve resource conflicts can lead to subtle performance problems. Consider the scenario illustrated in Figure 9-15. Tasks 1 and 2 each require access to a common resource protected by a semaphore. Task 1 has the highest priority and Task 2 has the lowest. Task 3, which has no need for the resource, has a "middle" priority.

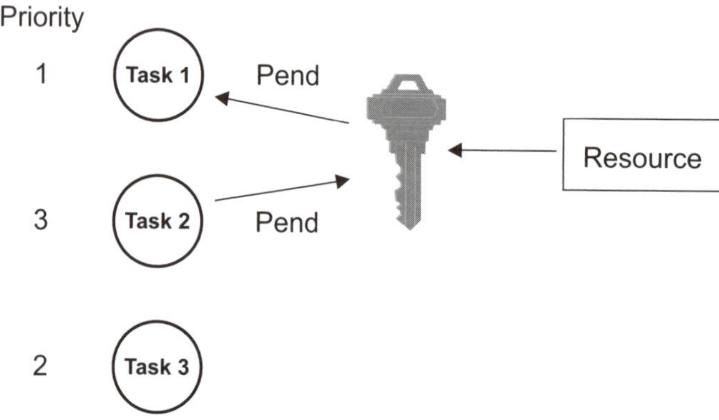

Figure 9-15: Priority Inversion Scenario

Figure 9-16 is an execution timeline of this system. Assume Task 2 is currently executing and pends on the semaphore. The resource is free so Task 2 gets it. Next an interrupt occurs that makes Task 1 ready. Since Task 1 has higher priority, it preempts Task 2 and executes until it pends on the resource semaphore.

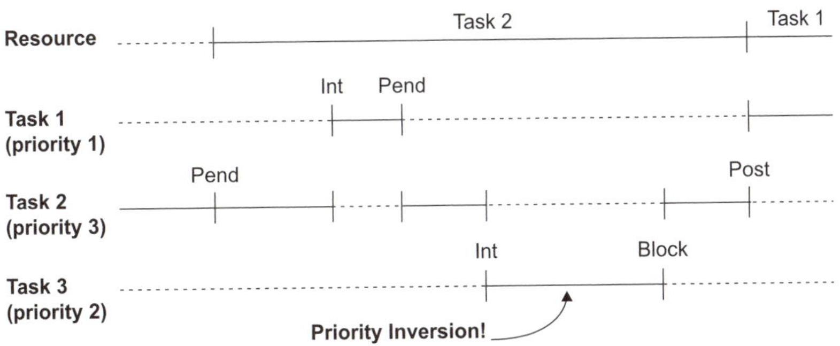

Figure 9-16: Priority Inversion Timeline

Since the resource is held by Task 2, Task 1 blocks and Task 2 regains control. So far everything is working as we would expect. Even though Task 1 has higher priority, it simply has to wait until Task 2 is finished with the resource.

The problem arises if Task 3 should become ready while Task 2 has the resource locked. Task 3 preempts Task 2. This situation is called *priority inversion* because a lower priority task (Task 3) is effectively preventing a higher priority task (Task 1) from executing.

A common solution to this problem is to temporarily raise the priority of Task 2 to that of Task 1 as soon as Task 1 pends on the semaphore. Now Task 2 can't be preempted by anything of lower priority than Task 1. This is called *priority inheritance*. If a kernel makes a distinction between a semaphore and a mutex, the latter will usually incorporate priority inheritance as an optional configuration parameter when the mutex is created.

Another approach, called *priority ceiling*, raises the priority of Task 2 to a specified value higher than that of any task that may pend on the mutex as soon as Task 2 gets the mutex. This is considered to be more efficient because it eliminates unnecessary context switches. No task needing the resource can preempt the task currently holding it.

Interrupts and Exceptions

Interrupt handling in the context of a multitasking kernel requires some special considerations. The problem is that the ISR may need to call a system service to notify a task that something has happened. This in turn could cause the scheduler to be invoked. There are two considerations here:

- The ISR must not call any service that would cause a task to block, i.e., MbxPend.
- Generally, the scheduler can't be invoked directly from within an ISR because the system is in a different *context*.

Fundamentally, this means the kernel needs to know when it's running in ISR context rather than task context.

There are two approaches to managing interrupts. I choose to call them the "direct" and "indirect" methods. The top half of Figure 9-17 shows the direct method where the interrupt vector points directly at your ISR. You create this connection with a system service such as SetVect, which places the address of your ISR at the appropriate vector location.

Note the keyword interrupt in the declaration of the interrupt handler function. This turns an ordinary C function into an ISR. It does two things:

- Save all registers on entry to the function and restore them before exiting.
- Replace the normal RET subroutine return instruction with an IRET interrupt return instruction.

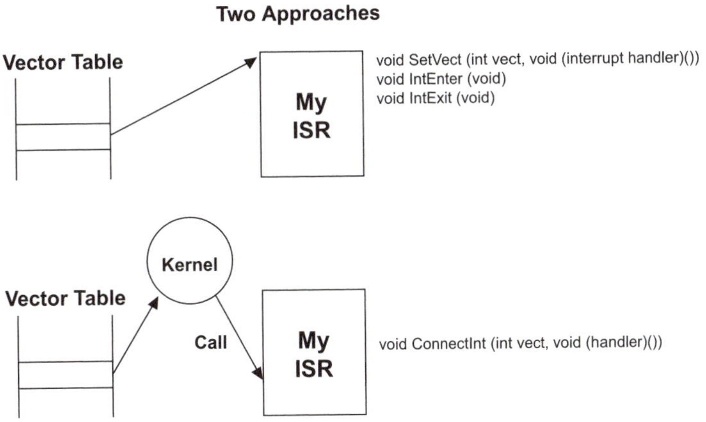

Two Approaches

Vector Table — My ISR

void SetVect (int vect, void (interrupt handler)())
void IntEnter (void)
void IntExit (void)

Vector Table — Kernel — Call — My ISR

void ConnectInt (int vect, void (handler)())

Figure 9-17: Interrupt Management

Your ISR is responsible for notifying the kernel that the system is operating in interrupt context. The functions IntEnter() and IntExit() provide for this notification.

The indirect method vectors the interrupt into the kernel which recognizes that it's entering interrupt context and performs the actions equivalent to IntEnter(). The kernel then *calls* your handler as an ordinary function. When the handler returns, the kernel performs the equivalent of IntExit() and returns from the interrupt. In this case, the handler needn't do anything with respect to the kernel and indeed doesn't even know it's running in the context of a kernel. Linux uses the indirect method.

Critical Sections

Within the operating system there are sections of code that must execute "atomically," that is they must not be interrupted under any circumstance. These are called *critical sections* and must be protected in much the same way as a printer is protected from multiple accesses by a semaphore. Whenever the kernel does a test and set operation on a global variable or updates a global linked list, this is a critical section.

A critical section is typically bracketed by a pair of functions of the form:

```
EnterCritical()
ExitCritical()
```

Code appearing between these functions is guaranteed not to be interrupted. In practice, these functions are often implemented as in-line assembly language macros that evaluate respectively to:

```
Disable interrupts
Enable interrupts
```

The performance of a real-time operating system is often characterized in terms of its interrupt latency, the maximum time from when an external interrupt is asserted until the ISR begins executing. Interrupt latency is largely determined by the kernel's critical sections. That is, how long does the kernel leave interrupts disabled? Consequently, kernel developers devote considerable effort to keeping the critical sections as short as possible.

It is sometimes useful to implement critical sections in application tasks. Suppose two tasks share access to a global 32-bit counter where one task reads the counter and the other increments it. Incrementing the counter must be done atomically, but on an 8- or 16-bit processor incrementing a 32-bit memory variable requires multiple instructions.

We could use a semaphore to protect the counter, but the overhead of the semaphore operations is probably orders of magnitude greater than the time needed to increment the counter. The more expedient solution is to simply:

```
EnterCritical();
        counter++;
ExitCritical();
```

As long as your critical sections are no longer than the longest critical section in the operating system, then your code is not affecting maximum latency.

Resources

Outside of Linux, the best resource by far for learning about and experimenting with real-time multitasking is MicroC/OS, a small preemptive multitasking kernel written almost entirely in C. MicroC/OS stands for Micro Controller Operating System and is designed to run on small 8- and 16-bit microcontrollers and has been ported to many of the most popular. It is thoroughly described in: Labrosse, Jean J., *MicroC/OS-II,The Real-time Kernel, 2nd Ed.*, 2002, CMP Books.

The book includes a CD with complete source code for MicroC/OS along with several examples that run in a DOS window on a PC. The code is extremely well written and can serve as a model of readability for all programmers. MicroC/OS is used in a number of commercial products and has been certified for use in safety-critical systems under the requirements of RTCA DO-178B.

Note that MicroC/OS is not Open Source and the code is not available for download. The only way to get it is to buy the book. Commercial use requires purchasing a license.

Other resources related to MicroC/OS are:

www.ucos-ii.com.

Simon, David E., *An Embedded Software Primer*, 1999, Addison-Wesley. David's book uses MicroC/OS as a tool in describing the embedded software development process.

Another interesting project is eCos, an Open Source, royalty-free, RTOS originally developed by Cygnus Solutions before it was acquired by Red Hat. eCos is highly configurable and supports a wide range of architectures and platforms in the 16- to 64-bit realm. To learn more about eCos, visit:

http://sources.redhat.com/ecos/

http://www.ecoscentric.com/

CHAPTER 10

Linux and Real-Time

"You can put racing stripes on a bulldozer but it won't go any faster."
—Victor Yodaiken

Why Linux Isn't Real-Time

Linux was conceived and built as a general-purpose multiuser operating system in the model of Unix. The goals of a multiuser system are generally in conflict with the goals of real-time operation. General purpose operating systems are tuned to maximize average throughput even at the expense of latency while real-time operating systems attempt to minimize, and place an upper bound on, latency, sometimes at the expense of average throughput.

There are several reasons why standard Linux is not considered suitable for real-time use:

- *Coarse-Grained Synchronization* – This is a fancy way of saying that kernel system calls are not preemptible. Once a process enters the kernel, it can't be preempted until it's ready to exit the kernel. If an event occurs while the kernel is executing, the process waiting for that event can't be scheduled until the currently executing process exits the kernel. Some kernel calls, `fork()` for example, can hold off preemption for tens of milliseconds.
- *Paging* – The process of swapping pages in and out of virtual memory is, for all practical purposes, unbounded. We have no way of knowing how long it will take to get a page off of a disk drive and so we simply can't place an upper bound on the time a process may be delayed due to a page fault.
- *"Fairness" in Scheduling* – Reflecting its Unix heritage as a multiuser time-sharing system, the conventional Linux scheduler does its best to be fair to all

processes. Thus the scheduler may give the processor to a low priority process that has been waiting a long time even though a higher priority process is ready to run.

- *Request Reordering* – Linux reorders I/O requests from multiple processes to make more efficient use of hardware. For example, hard disk block reads from a lower priority process may be given precedence over read requests from a higher priority process in order to minimize disk head movement or improve chances of error recovery.
- *Batching* – Linux will batch operations to make more efficient use of resources. For example, instead of freeing one page at a time when memory gets tight, Linux will run through the list of pages clearing out as many as possible, delaying the execution of all processes.

You have probably already noticed the consequences of these issues in using Linux or even Windows on your PC. Try moving the mouse while executing a compute-intensive function like rendering a complex graphics image, or while connecting to a dial-up line. The mouse occasionally stops and then jumps because the compute- or I/O bound process has the CPU locked up. In a desktop environment this is nothing more than irritating. In a real-time environment it's unacceptable and may even be catastrophic.

The net effect of all these characteristics is that we can't put an upper bound on the latency that a user task or process may encounter. By definition this is not real-time.

Measuring Latency—An Experiment

Let's try an experiment to see just how much latency varies in standard Linux. Start by untarring the book CD file `Rtdemos.tar.gz` into your home directory. You now have a directory named `Rtdemos/` and under it is a directory named `ProcessJitter/`. `cd` to `ProcessJitter/` and `open jitter.c`.

Move down to around line 60. Jitter uses the `select()` system call to sleep for less than a second, in this case 50 milliseconds. The normal use of `select()` is to wait for any of a specified set of file descriptors to change status. Here we're simply using the timeout argument.

When `jitter` wakes up it reads the current time of day and computes the interval from the last call to `gettimeofday()`. Next it computes, in microseconds, the deviation between the actual time interval and the expected interval of 50 milliseconds. Finally, it updates and prints minimum, maximum, average, and current deviations.

Make jitter and start it running. On my lab workstation running X windows I see a maximum variance in the range of about four to five milliseconds with nothing much going on. The average is on the order of 100 microseconds. Try some simple things such as changing directories in a file manager window or opening a text file with the editor. You should see the maximum variance go up.

Now introduce a *real* load on the system. The easiest way to do that is to start up Netscape. You should see maximum variance jump up into the tens of milliseconds. There you have it. Linux is not real-time. To terminate jitter, just hit <Enter>.

Incidentally, this experiment assumes your 2.6.x kernel does not have preemption enabled.

Improving Linux Latency

There are some things we can do to improve the latency of standard Linux. Specifically, we can change the kernel's "scheduling policy" and process priority for the jitter process and we can lock the process's memory image into RAM so it won't be paged out.

The default scheduling policy, called *SCHED_OTHER*, uses a fairness algorithm and gives all processes using this policy priority 0, the lowest priority. This is fine for "normal" processes. The alternate scheduling policies are SCHED_FIFO and SCHED_RR. These are intended for time-critical processes requiring lower latency. Processes using these alternate scheduling policies must have a priority greater than 0. Thus a process scheduled with SCHED_FIFO or SCHED_RR will preempt any running normal process when it becomes ready.

Conceptually, the kernel maintains a FIFO queue of runnable processes for each possible priority value ranging from 0 to 99. To determine the next process to run, the scheduler finds the highest priority nonempty queue and takes the first entry. If a SCHED_FIFO process is preempted by a higher priority process it remains at the head of its priority queue. When a SCHED_FIFO process becomes runnable after being blocked, it goes at the back of the queue.

A SCHED_FIFO process runs until it blocks or it yields. SCHED_RR is a minor variation on SCHED_FIFO that adds time slicing. If a SCHED_RR process exceeds its time slice it is placed at the back of its priority queue.

Now look at the section of jitter.c beginning at line 27. If you run jitter with an argument "high", it sets the scheduling policy with a call to sched_setscheduler() and then locks the process in memory with a call to mlockall().

You will need to be super user to run this version of jitter. If you're not already root, enter the su command and root password. Now execute jitter high and start up Netscape again. The maximum variance should be no more than around half of what you saw before. Better, but still not real-time.

Two Approaches

OK, so Linux is not real-time. What do we do about it? Well, there are at least two very different approaches to giving Linux deterministic behavior.

Preemption Improvement

One approach is to modify the Linux kernel itself to make it more responsive. This primarily involves introducing additional preemption points in the kernel to reduce latency. An easy way to do this is to make use of the "spinlock" macros that already exist in the kernel to support symmetric multiprocessing (SMP). In an SMP environment spinlocks prevent multiple processors from simultaneously executing a critical section of code. In a uniprocessor environment the spinlocks are no ops.

The preemption improvement strategy turns the spinlocks into the equivalent of EnterCritical() and ExitCritical() that we encountered in the last chapter. So whereas the standard kernel prevents preemption unless it's specifically allowed, the preemptible kernel allows preemption unless it's specifically blocked by a critical code section.

Interrupt handling is also modified to allow for rescheduling on return from an interrupt if a higher priority process has been made ready. This approach is often coupled with a new scheduler that provides fixed overhead for real-time tasks. Historically, Monta Vista and TimeSys have been the principal proponents of preemption improvement.

Note by the way that when we speak of kernel preemption, we're referring to *process* latency, not *interrupt* latency. With a standard uniprocessor Linux kernel, interrupt latency is on the order of 60 microseconds, depending of course on processor speed. As noted above, maximum process latency for a standard kernel is in the tens of milliseconds. The preemption improvement strategy reduces that to 1 to 2 milliseconds.

The advantage to the preemption improvement approach is that the real-time applications run in user space just like any Linux application using the familiar Linux/Posix APIs. Real-time processes are subject to the same memory protection rules as ordinary processes. But it's still not hard real-time. Latency is reduced but there are simply too many execution paths in the kernel to permit comprehensive analysis of determinism.

For 2.4 series kernels, the preemption improvements are available as a kernel patch. The 2.6 series merges the preemption patch into the main kernel development tree as a configuration option. The "Preemptible kernel" option shows up under "Processor type and features" in the kernel configuration menu[1].

Along with kernel preemption, the process scheduler in the 2.6 kernel received a major overhaul. The previous scheduler exhibited what is called O(n) behavior meaning that the time required to schedule the next process was proportional to the number of ready processes in the queue. That is, the scheduler traversed the entire queue to determine which process had the highest claim on the processor at this instant in time. The new scheduler exhibits O(1) behavior meaning that the scheduling time is independent of the number of ready processes.

So-called "big-O" notation is a way of representing the time complexity of algorithms as a function of the number of inputs. Many algorithms exhibit O(n) behavior while others can be much worse, as in $O(n^2)$ or, really bad, O(n!). Most of the time, the best behavior is O(1), meaning that the execution time is constant with respect to the number of inputs. But, on the other hand, if the constant time is three hours, it might be better to look at a O(n) algorithm that may actually run faster for the most likely number of inputs.

Rebuild your kernel with preemption enabled, run `jitter` and see what kind of numbers you get now for maximum latency.

Before moving on to the other approach to real-time Linux, we should take a quick look at another configuration option that surfaced in the 2.6 kernel. You now have the option to turn off virtual memory. No more paging! This addresses the second major impediment to real-time behavior in Linux listed at the beginning of the chapter.

Turning off virtual memory is fine as long as you can anticipate the maximum amount of memory your system will require. In fact, in an embedded situation this is often possible. You know what applications will be running and how much memory they require. So just make sure that much RAM is available.

The option to turn off virtual memory is called "Support for paging of anonymous memory" and shows up under "General setup" in the configuration menu.

---

[1] It's not clear to me why this is an option. Under what circumstances would you not want a preemptible kernel?

Interrupt Abstraction

The improvements in the 2.6 kernel, while significant, still don't get us to deterministic, real-time performance. Even though the default scheduler executes in fixed time, it still attempts to be "fair". There's also the problem of the block I/O system reordering and combining I/O requests in the interest of maximizing throughput. So we need to circumvent these issues, at least for the parts of the application that are truly real-time.

It turns out that in a great many applications, only a small part of the system truly requires hard real-time determinism. Controlling a high-speed PID loop or moving a robot arm are examples of hard real-time requirements. But logging the temperature the PID loop is trying to maintain, or graphically displaying the current position of the robot arm are generally not real-time requirements.

The alternate, and some would say more expedient, approach to real-time performance in Linux relies on this distinction between what is real-time and what is not. The technique is to run Linux as the lowest priority task (the idle task if you will) under a small real-time kernel. The real-time functions are handled by higher priority tasks running under this kernel. The non real-time stuff, such as graphics, file management and networking, which Linux already does very well, is handled by Linux.

This approach is called "Interrupt Abstraction" because the real-time kernel takes over interrupt handling from Linux. The Linux kernel "thinks" it's disabling interrupts but it really isn't. The essence of interrupt abstraction is illustrated in Figure 10-1. The real-time kernel effectively intercepts hardware interrupts before they get to the Linux kernel. Linux no longer has direct control over enabling and disabling interrupts. So when Linux says disable interrupts, the RT kernel simply clears an internal software interrupt enable flag but leaves interrupts enabled. When a hardware interrupt occurs, the RT kernel first determines to whom it is directed:

- *RT Task*: Schedule the task
- *Linux*: Check the software interrupt flag. If enabled, invoke the appropriate Linux interrupt handler. If disabled, note that the interrupt occurred and deliver it later when Linux re-enables interrupts.

The Linux kernel is treated as the lowest priority, or idle, task under the RT kernel and only runs when there are no real-time tasks ready to run.

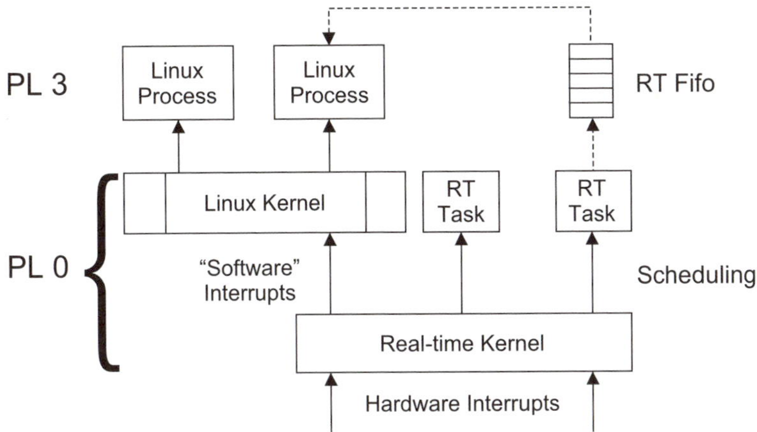

Figure 10-1: Interrupt Abstraction Architecture

Of course, there will be times when the RT kernel has to disable hardware interrupts to manage its own critical sections, but these are of much shorter duration than the critical sections in Linux.

The real-time tasks will usually have some need to communicate with user-space processes for things such as file access, network communication or user interface. The RT kernel provides mechanisms such as FIFOs and shared memory that support communication with user space processes.

Being much smaller and simpler, the real-time OS is amenable to execution time analysis that provides reliable upper bounds on latency. And while this approach also involves modifying the kernel, the extent of the modifications is substantially less than the Preemption Improvement approach.

The Interrupt Abstraction RTOS introduces its own API and purists insist that this is not "true" Linux. But as the interrupt abstraction approach has evolved over the years, the two major implementations have both evolved "wrappers" around the native API for Posix threads.[2] This mitigates, to some extent, the objections of the purists. This also means that real-time tasks can be initially tested in User space using the standard Posix threads library and conventional debug tools before moving into the real-time environment.

The real-time tasks run in kernel space.[3] This is a good-news, bad-news situation. The good news is that response times in kernel space are very short. Interrupt re-

[2] We'll look at POSIX threads in Chapter 12.

[3] RTAI does in fact offer real-time functionality in user space. We'll see that in the next chapter.

sponse and task switching times under 10 microseconds are the norm. The bad news of course is that there's no protection in kernel space and a real-time task can bring down the whole system. The real-time tasks are in effect extensions of the kernel.

There are two major implementations of the Interrupt Abstraction approach:

- *RTLinux*: This is the original interrupt abstraction implementation. It was developed at the New Mexico Institute of Mining and Technology under the direction of Victor Yodaiken. While an Open Source version of RTLinux is still available, much of the development work is going into a proprietary version called RTLinux/Pro offered by FSM Labs, Inc.
- *RTAI*: This is an enhancement of RT Linux developed at the Dipartimento di Ingeneria Aerospaziale, Politecnico di Milano under the direction of Prof. Paolo Mantegazza. It is a very active Open Source project with many contributors.

A Third Way—the Nanokernel

There's a problem for the Open Source community with the interrupt abstraction method. Victor Yodaiken, who founded FSM Labs to commercialize RTLinux, has a patent[4] on the technique. The licensing terms are somewhat complicated. Among other provisions, the patented technique is available for free use provided:

1. you use RTLinux unmodified, that is "out of the box," or;
2. you use the technique only with software released under the GPL.

The second provision seems directed specifically at RTAI, which is released under the LGPL rather than the GPL. As a result, RTAI has migrated to ADEOS, the Adaptive Domain Environment for Operating Systems.

Adeos is a "nanokernel" that runs underneath one or more operating systems, allowing them to share hardware resources. Each OS is encompassed in a "domain" over which it has total control and, indeed, the OS need not even be aware of Adeos. The domain may include a private address space and software abstractions such as processes, virtual memory, file systems, and so on.

Of necessity, Adeos must take control of some system hardware to accomplish its task, but it tries not to intrude unnecessarily on the operation of the respective OSes. The basic architecture is illustrated in Figure 10-2, which shows the four categories of communication relevant to an Adeos system. Category A represents normal memory

[4] U.S. Patent 5,995,745, "Adding Real-Time Support to General-Purpose Operating Systems."

and I/O access by an operating system independent of Adeos. At this level, each domain operates as if Adeos weren't even there.

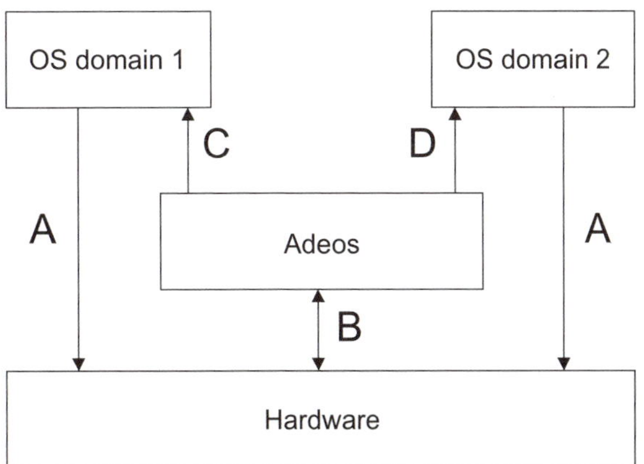

Figure 10-2: Adeos Architecture

Category B represents Adeos receiving control from the hardware as the result of a hardware or software interrupt. It also includes commands issued by Adeos to control the hardware as necessary. Adeos uses the Category C line to invoke the interrupt handler of an OS. If the target OS is not Adeos aware, this may involve setting up the stack to look like what the interrupt handler expects. Category D represents two-way communication that can occur with a domain that is Adeos aware. Such domains can use the facilities of Adeos to share resources.

For all practical purposes, Adeos is a form of interrupt abstraction, but it is implemented in a way that gets around the claims made in the 745 patent. Figure 10-3 shows how Adeos handles interrupts. Adeos propagates interrupts down a "pipeline" populated by the various domains running in the system. Domains can be ordered in the pipeline according to how critical it is that they get hardware interrupts first.

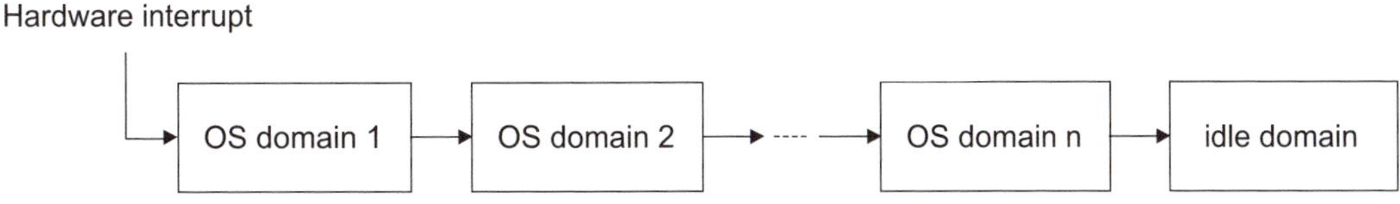

Figure 10-3: Adeos Interrupt Pipeline

Each domain has the option to accept, ignore, discard, or terminate interrupts. If a domain is accepting interrupts, the normal case, Adeos sets up and invokes its interrupt handler. The OS in the domain can do whatever it wants in response to the interrupt. Eventually the OS reaches its idle task, at which point Adeos is notified to pass the interrupt on to the next stage in the pipeline.

A domain can choose to "ignore" interrupts, which is the same thing as disabling interrupts. The effect in Adeos is to stall the pipeline at that stage. Interrupts proceed no further down the pipeline until the domain decides to accept them again. A domain that chooses to "discard" interrupts is simply passed over and the interrupts continue down the pipeline. If a domain "terminates" interrupts, they are not propagated beyond that stage of the pipeline. Only Adeos-aware domains can discard or terminate interrupts.

Getting Adeos to manage OS domains that are not aware of its existence is a cute trick. On an x86, the Linux kernel normally runs with the highest privileges, at Privilege Level (PL) 0. Adeos, running as a kernel module, modifies the Global Descriptor Tables to move the Linux kernel down to PL 1 while Adeos itself continues to run at PL 0. The upshot is that privileged instructions executed by Linux cause processor traps that are intercepted and dealt with by Adeos.

Note by the way that it's not necessary to understand Adeos in any detail to use RTAI. Nevertheless, it's an interesting technology and worth additional investigation if you're so inclined.

Resources

Obtaining Real-Time Linux Implementations

The book CD includes RTAI version 3.1 that is compatible with version 2.6 kernels. We'll explore that in detail in Chapter 11. The latest version of RTAI is available at *www.rtai.org*. Here is a short sampling of other real-time Linux resources:

The Open Source version of RTLinux is available at *http://www.rtlinuxfree.com/*.

The Real Time Linux Foundation is a nonprofit corporation whose charter is to support the real-time Linux community. Visit them at *http://www.realtimelinux-foundation.org/*.

Monta Vista Software offers several versions of Linux with real-time characteristics including the "Professional Edition," "Carrier Grade Edition," "Consumer Electronics Edition," and "Mobilinux" for mobile devices. Further information is available at *www.mvista.com*.

LynuxWorks offers a Linux ABI-compatible RTOS called LynxOS. More information can be found at: *www.bluecat.com*.

Additional information on Adeos is available from the project website at *www.gna.org/adeos*.

CHAPTER 11

The RTAI Environment

In this chapter we'll explore the RTAI environment running on the workstation. In principle, RTAI should run on your target with BlueCat Linux, since it's also x86 based, however, the version of BlueCat we have is based on the version 2.6.0 kernel but RTAI 3.1 only has patches for versions 2.6.7 and 2.6.8 plus a few later versions of the 2.4 kernel. If you really want to try RTAI on your target, go to the RTAI download page and find a version that supports the 2.6.0 kernel.

RTAI has a rather extensive set of APIs. Rather than try to describe all of them in detail, we will focus on some examples that illustrate the basics. The complete APIs are listed in Appendix A.

Installing RTAI

Under the directory /RTAI on the book CD are two tar files:

- rtai-3.1-tar.bz2 — Untar to /usr/src
- Rtdemos.tar.gz — We untarred this one in the previous chapter

Create the symbolic link:

```
rtai -> rtai-3.1
```

Patching the Kernel

RTAI requires changes to the base Linux kernel source code to add support for the Adeos nanokernel. Linux needs to be aware of Adeos to support communication between real-time tasks running in Kernel space and normal User space processes.

The mechanism for changing released source code in an orderly manner is the ***patch*** utility. The input to patch is a text file created by the ***diff*** utility that compares two files and reports any differences. So when an Open Source programmer wants to distribute an upgrade to released source code, she does a diff between the modified code and the original code redirecting the output of diff to a file.

The resulting file is called a *patch file*. When you want to implement the modifications on your system, you start with the same original source code and apply the patch. Distributing patch files is generally much more efficient than distributing the entire modified source tree.

Take a look at rtai/rtai-core/arch/i386/patches/. It contains patch files for several x86 kernel versions. Other subdirectories under arch/ provide patches for other ports including ARM, MIPS, and Power PC. We'll be using hal7-2.6.7.patch to patch the 2.6.7 kernel. Take a look at that file just to get a feel for what a patch file looks like.

Before patching the kernel source tree, it's a good idea to make a copy. Suppose the patch fails. You could be left with a partially patched source tree that may not build. It's always nice to have a clean source tree to fall back on. So copy /usr/src/linux-2.6.7 and all its subdirectories to /usr/src/linux-2.6.7-adeos. You'll be copying about 17,000 files, so this would be a good time to take a break.

Redirect the symbolic link linux in /usr/src to point to linux-2.6.7-adeos/ and cd /usr/src/linux. Now execute the command

```
patch –p1 < ../rtai/rtai-core/arch/i386/patches/ hal7-2.6.7.patch
```

patch will list the files it patched.

The –p1 flag tells patch to remove one slash and everything before it from the names of files to be patched. The file names in hal7-2.6.7.patch all begin with "linux-2.6.7/" or "linux-2.6.7-adeos/". Since we're already in linux-2.6.7-adeos/, we use –p1 to remove the unnecessary directory information.

Although Adeos claims that the operating systems that run on top of the nanokernel need not be aware of its existence, it modifies a relatively large number of files in the Linux kernel. More, in fact than the original Hardware Abstraction Layer approach to RTAI, which only modified 21 files. For the record, Listing 11-1 lists the files modified and created to support Adeos.

```
Makefile                                 include/linux/init_task.h
init/main.c                              include/linux/preempt.h
                                         include/linux/sched.h
kernel/Makefile                          include/asm-i386/smp.h
kernel/exit.c                            include/asm-i386/system.h
kernel/fork.c                            include/asm-i386/io_apic.h
kernel/panic.c                           include/asm-i386/mach-default/
kernel/printk.c                            do_timer.h
kernel/sched.c                           include/asm-i386/mach-default/
kernel/signal.c                            irq_vectors.h
kernel/sysctl.c                          include/asm-i386/mach-pc9800/
                                           do_timer.h
arch/i386/Kconfig                        include/asm-i386/mach-pc9800/
arch/i386/kernel/Makefile                  irq_vectors.h
arch/i386/kernel/apic.c                  include/asm-i386/mach-visws/
arch/i386/kernel/cpu/mcheck/p4.c           do_timer.h
arch/i386/kernel/entry.S                 include/asm-i386/mach-visws/
arch/i386/kernel/i386_ksyms.c              irq_vectors.h
arch/i386/kernel/i8259.c
arch/i386/kernel/io_apic.c               New Files
arch/i386/kernel/irq.c
arch/i386/kernel/nmi.c                   Documentation/adeos.txt
arch/i386/kernel/process.c               adeos/Kconfig
arch/i386/kernel/signal.c                adeos/Makefile
arch/i386/kernel/smp.c                   adeos/generic.c
arch/i386/kernel/smpboot.c               adeos/x86.c
arch/i386/kernel/time.c                  kernel/adeos.c
arch/i386/kernel/timers/timer_pit.c      arch/i386/kernel/adeos.c
arch/i386/kernel/timers/timer_tsc.c      include/asm-i386/adeos.h
arch/i386/kernel/traps.c                 include/linux/adeos.h
arch/i386/mm/fault.c
```

Now we need to configure and build our newly patched kernel. Do make xconfig. Down near the bottom of the left hand navigation panel is a new entry called *Adeos support*. It has one option, also called *Adeos support*, that should be selected. You'll also find a new subdirectory under linux-2.6.7-adeos/ called, perhaps not surprisingly, adeos/. Under "Loadable module support" we need to turn off "Module versioning support." This avoids missing links when loading RTAI modules.

The default processor type for an x86 kernel is usually 386 so that the resulting kernel can run on just about anything. However, RTAI insists on having better than a 486 to make use of features that first appeared in Pentium processors. So under

"Processor type and features," you'll need to select (and of course have available) a 586 or better processor.

Make sure that other configuration options are correct for your workstation and then build the kernel as described in Chapter 4. Copy the resulting bzImage file to /boot with the name vmlinuz-2.6.7-adeos. Copy System.map as well. Update lilo.conf or grub.conf as necessary to add the new kernel. Note that you can use the same initrd file that you made back in Chapter 4 for the base 2.6.7 kernel.

After booting your new "adeos-enabled" kernel, you'll find a new file in the /proc directory, adeos. This provides some cryptic information about the Linux domain running under Adeos.

Configuring and Building RTAI

RTAI uses the same configuration mechanism as the kernel. cd /usr/src/rtai and do make xconfig. Eventually the menu in Figure 11-1 will appear. Scroll through the menu to see what the options are. For now we'll stick with the defaults. When you're finished browsing close the menu and save the config file.

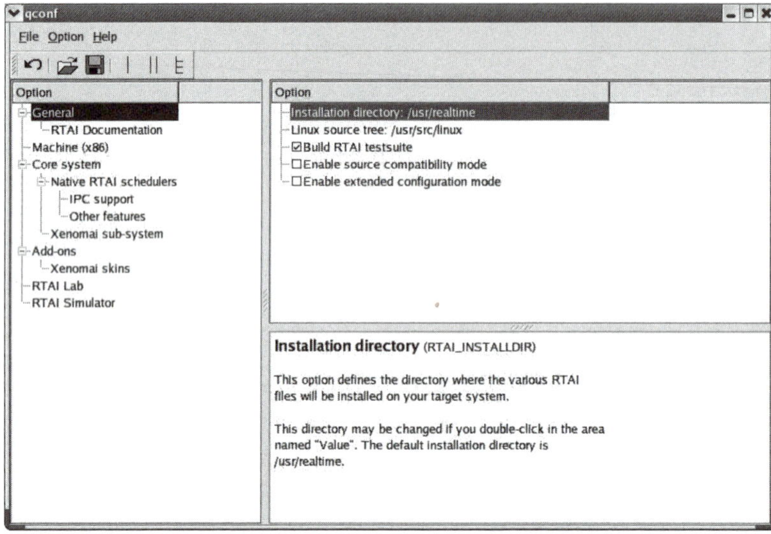

Figure 11-1: RTAI xconfig Menu

RTAI expects to put the result of its build process into an *installation directory* that is separate from the source tree. The default, as you may have noticed in the xconfig menu, is /usr/realtime. That seems like as good a place as any, so go ahead and create a directory realtime/ under usr/.

To build RTAI, execute the following make commands:

make	Builds the RTAI modules (takes a while). The first time you build RTAI, make begins with the menuconfig configuration menu. Just exit out without saving. RTAI uses a Linux build feature called *autoconf*. Before actually starting a build, autoconf spends a fair amount of time checking various features of the system to try and make sure the build is likely to succeed.
make install	Installs the modules and a number of other files in the default installation directory, /usr/realtime

When the build installation is complete, /usr/realtime/modules contains all the RTAI kernel modules[1]. There are quite a few of them but for our purposes the most important are:

rtai_hal	The RTAI Adeos interface layer
rtai_ksched	A link to the scheduler, in this case ...
rtai_up	The uni-processor scheduler
rtai_mbx	Mailboxes
rtai_msg	Messaging
rtai_sem	Semaphores
rtai_fifos	Real-time FIFOs
rtai_shm	Shared memory
rtai_lxrt	User Space real-time tasks

rtai_ksched uses services provided by rtai_hal. In turn, the other modules use services provided by rtai_ksched. So there is a definite order to the loading of RTAI modules. rtai_hal must be loaded first, rtai_ksched next. The remaining modules don't depend on each other and can be loaded in any order.

---

[1] This is a change from previous releases of RTAI and, in my opinion, not necessarily an improvement. Previously, the RTAI modules were put into a subdirectory of /lib/modules/<kernel_version> and were added to the module dependency list where modprobe could find them. modprobe can't find them where they are now. The RTAI developers have instead created a somewhat cryptic load script strategy that automates the process of loading and unloading an RTAI application.

As the Superuser, cd to /usr/realtime/modules and try the following:

```
/sbin/insmod rtai_hal.ko
/sbin/insmod rtai_ksched.ko
/sbin/lsmod
cat /proc/rtai/rtai
cat /proc/rtai/scheduler
```

Many RTAI modules create /proc files so you can see what's going on.

The directory /usr/realtime/bin contains some scripts that will be useful later on, so add that to your PATH.

Inter-Task Communication and Synchronization

RTAI includes an extensive set of mechanisms to facilitate communication and synchronization among Kernel Space tasks. These include:

- *Semaphores*. Conventional counting semaphores as described in Chapter 9. The semaphore operations are signal and wait.
- *Mailboxes*. RTAI mailboxes are much like the queues described in Chapter 9. The mailbox is created with arbitrary size and arbitrary quantities of data can be written to and read from a mailbox.
- *Messaging*. This is a direct task-to-task communication mechanism. A task can send a single integer directly to another task. The receiving task can wait for a message (integer) from a specific sender or from any task. There is also a full duplex version of messaging called *Remote Procedure Calls*.

Communicating with Linux Processes

The communication and synchronization mechanisms described above apply only to Kernel Space real-time tasks. RTAI also supports two mechanisms to provide communication between a real-time task and a User Space Linux process. These are RT FIFOs and shared memory.

RT FIFOs

A RT FIFO is a point-to-point link connecting one real-time task to one Linux process. It's very much like a Unix pipe. The implementation allows a FIFO to be bidirectional, but in practice that rarely makes sense. Suppose for example that one end of the FIFO writes a command and then immediately tries to read the result of the command it just wrote. Chances are it would just read back the command it wrote. So in practice, FIFOs are unidirectional where the direction is established by the programmer. In the example just cited, you would create two FIFOs; one to send the command and the other to read the response.

User Space processes treat RT FIFOs as character devices, `/dev/rft0` to `/dev/rtf63`. A process opens a FIFO for reading or writing and then uses `read()` or `write()` on the file descriptor to transfer data. The rtf nodes are automatically added to your filesystem by the RTAI build process if you select FIFO support.

Real-time tasks access the FIFO through an RTAI-specific API.

We'll look at the FIFO API in more detail later when we revisit the thermostat example.

Shared Memory

The FIFO model is useful in situations where a relatively small amount of data must be transferred between one RT task and one User Space process more or less synchronously. But there are also situations where multiple processes might require access to data generated by a single real-time task. In this case a shared memory model makes more sense. Remember that we used shared memory back in Chapter 8 for the networked thermostat.

There are also situations where large amounts of data, video frame buffers for example, must be moved quickly between a real-time task in Kernel Space and one or more processes in User Space. Here's another case where shared memory makes sense because the data doesn't have to be copied from one domain to the other. The real-time task writes the shared memory region and the processes read it at their leisure.

Real-Time in User Space—LXRT

RTAI and the real-time tasks that it manages run in Kernel Space at Privilege Level 0. As we've seen before, this leads to problems during development because it's difficult to use a source level debugger such as GDB on kernel code. Fortunately RTAI has an escape hatch called *LXRT*. LXRT allows you to run real-time tasks in User Space using the same API that is provided in Kernel Space RTAI.

How it Works

You create an LXRT task as an ordinary Linux User Space process with a main() function. As part of the initialization, you create a "buddy" task that operates in Kernel Space on behalf of the User Space task. When, for example, your LXRT task calls rt_task_wait_period(), LXRT gets your buddy, running in Kernel Space, to execute the real function. Control returns to the LXRT task only when its buddy wakes up. Similarly, you can create RTAI communication and synchronization objects such as semaphores and mailboxes. Each of these objects is identified by a name. You pass the object's name to the appropriate initialization or creation function, which returns a pointer to a data structure to be used in accessing the object.

Measuring Latency with LXRT

As an introduction to RTAI we'll run a variation on the latency experiment of the previous chapter. This version is structured as a pair of LXRT tasks. One task, called rt_process, wakes up periodically, reads the current time, computes the deviation from the ideal period, and checks min and max values. After some number of passes it sends this information via an RTAI mailbox to the other task, check, which simply displays the results. This is in fact a simple variation on one of the RTAI LXRT examples.

cd Rtdemos/TaskJitter and open rt_process.c. Scroll down into main() around line 69. For LXRT to function correctly, you must use the SCHED_FIFO scheduling policy and lock the process in memory (no paging). The function rt_task_init() creates an LXRT task named "LATCAL" and returns a pointer to a RT_TASK structure. This is a handle by which we'll refer to this task from now on. nam2num() is just a way of "mangling" up to six ASCII characters into a single integer ID. rt_task_init() leaves the task in the suspended state.

Next we create a mailbox named "LATMBX" with a call to rt_mbx_init(), which returns a pointer to a mailbox structure. Within the if statement concerning oneshot, the function start_rt_timer() starts the system timer with the specified period, in this case 100,000 nanoseconds, or 100 microseconds. Note that the argument to start_rt_timer() is in internal timer counts. For the time being skip over the other code in this if statement as well as the one concerning hardrt. We'll come back to these later.

LATCAL is a periodic task so we start it with a call to rt_task_make_periodic(). The task starts now with the same period as the timer, so it wakes up every timer tick.

Inside the task's while() loop it calls rt_task_wait_period() to block until its next scheduled execution time as specified by rt_task_make_periodic(). Each time the task wakes up it reads the current time and computes the deviation from the expected wakeup time. After a few thousand passes it sends a data structure to the mailbox with the function rt_mbx_send_if(). The "if" suffix means the message will be sent only if there's space for it in the mailbox. This is a nonblocking version of sending to a mailbox

rt_receive_if() tests to see if the companion task "LATCHK" has sent a message. A nonzero return indicates a message was received. In this case the message value is unimportant, it merely says that LATCHK is going away and so LATCAL should go away too. The function rt_get_adr() looks up a name in the LXRT name space and returns the corresponding handle if the name has been registered by an object initialization function.

Now take a look at check.c. It creates a buddy task for itself and then gets the address (handle) of LATMBX. Note we're assuming that LATCAL starts first. rt_mbx_receive() is a blocking call that doesn't return until it receives a message in the mailbox. Then it simply prints out the various deviation parameters. Finally it tests the flag set by the signal handler to see if it's time to go away. If so it sends a dummy message to jitter and returns. main() then deletes the task.

Let's try it out. Execute the following make commands:

```
make rt_process
make check
```

Although you can, and should, make these targets as a normal user, you will, of course, need to be Superuser to run the programs because only Superuser processes are allowed to set the scheduler and lock memory.

But before we can run either of these processes, we need to load a number of RTAI modules. There are a couple of ways to do this. The directory Rtdemos/ has a pair of script files called *ldmod* and *remod*. These load and remove, respectively, just about any RTAI module you might need. These scripts can be invoked from any of the subdirectories under Rtdemos/.

With the RTAI environment set up, we can now run our two tasks. There are two ways to do this. We could start two shell windows and run one program in each window. Or, we can run rt_process as a background task and run check as the foreground task in the same window. Enter:

```
./rt_process &
```

The "&" directs bash to spawn a process for rt_process but then immediately come back for another command rather than waiting for rt_process to finish. Now enter:

```
./check
```

The alternative for starting up everything is a clever shell script called rtai_load that resides in /usr/realtime/bin. It expects to find a file called .runinfo in the current directory that specifies a set of RTAI modules to load, and possibly some user space processes to run plus information to be communicated to the user.

The .runinfo file in TaskJitter/ looks like this:

```
latency:lxrt+sem+mbx+msg:./rt_process &;sleep 1;./check;popall:control-c
```

The single line is divided into four sections separated by colons. The first section is just a target name and really doesn't mean much unless the file has more than one target. The second section is a shorthand list of RTAI modules to load without the "rtai_" prefix and the ".ko" extension. The third section is a list of User Space processes that will be run in order. Note here that after starting rt_process in the background, we wait for a second to make sure it's started before starting check. The final entry in this section is a keyword to rtai_load that tells it to unload all the modules it loaded previously.

Finally, the last section provides a way to communicate with the user. control-c is a built-in message that tells the user to type control-C to terminate the program. More details on the .runinfo file are available in /usr/src/rtai/rtai-doc/maintainers/runinfo.txt.

The results of the latency check are displayed in nanoseconds. Fire up Netscape and watch what it does to the maximum deviation. It's pretty bad isn't it. In fact, it's not much better than what we saw in the last chapter with straight Linux. So conventional LXRT really is just "soft" real-time. It's useful as a debugging tool before you move your tasks into Kernel Space but it doesn't buy you anything in terms of deterministic performance.

When you're finished, type <Enter> into the window running check and it will terminate both processes.

Hard Real-Time in LXRT

But that's not the end of the story. Later releases of LXRT include a feature that allows you to get hard real-time performance in User Space. The function rt_make_hard_real_time() turns a User Space process into a hard real-time process (perhaps the more accurate term is hard*er*). It does this by playing games with the scheduler and blocking hardware interrupts while a hard real-time process is executing. You can return the process to normal soft real-time by calling rt_make_soft_real_time().

rt_process provides for a couple of run-time parameters, beginning respectively with "h" and "o", to alter its behavior. The "h" stands for hard real-time. If you execute ./rt_process h, the program invokes rt_make_hard_real_time(). Try it out. You should see substantially better results, down in the range of tens of microseconds or less even when starting up Netscape.

One Shot vs. Periodic Timing

RTAI supports two modes of timer tick interrupt, called *periodic* and *one-shot*. In periodic mode you set the timer to interrupt at a specified period. When the timer counter overflows and generates the interrupt, it is automatically reloaded with the correct starting number and starts over. Thus there is no overhead in servicing the timer chip itself.

In fact this is the way most operating systems work. The downside is that periodic tasks can only be scheduled in increments of the timer tick interval. If the tick interval is one millisecond for example, the shortest task period is one millisecond. Of course if you need finer granularity you can decrease the tick interval to, say 100 microseconds, or maybe even 10 microseconds. The problem here is that servicing the timer tick imposes some fixed overhead and the shorter the tick interval, the higher percentage of processor time is devoted to tick servicing. At a tick interval of 10 microseconds you would probably find the system almost entirely devoted to servicing the tick interrupt leaving little time to do real work.

RTAI's one-shot timing mode is a solution to this problem. Whereas periodic mode just lets the timer free run, interrupting at the specified interval, one-shot mode reprograms the timer each time it interrupts. That is to say at every tick interrupt, we compute the time to the next "event" and program the timer for that interval. The resolution of the timing interval is now determined by the clock driving the timer and not the periodic interval of timer interrupts.

Consider a simple example involving the following three periodic tasks:

Task	***Interval***
Task1	1.3 milliseconds
Task2	600 microseconds
Task3	2.1 milliseconds

Assume for the sake of simplicity that all three tasks are started simultaneously. They will be ordered on a waiting list with the shortest interval first and the other tasks expressed as time remaining after the first interval expires:

Task	***Remaining Interval (microseconds)***
Task2	600
Task1	700
Task3	1500

The timer is set for 600 microseconds. When the interrupt occurs, Task2 is made ready and the list is updated to reflect the next interval. In this case, Task2's 600 microsecond period is still less than the remaining intervals for the other tasks and so the timer is again set for 600 microseconds and the list is updated as follows:

Task	***Remaining Interval (microseconds)***
Task2	600
Task1	100
Task3	900

At the next timer interrupt Task2 is again made ready, the list is updated as follows and the timer is set for 100 microseconds:

Task	***Remaining Interval (microseconds)***
Task1	100
Task2	500
Task3	800

The trade-off for this flexibility is more timer overhead due to the need to recalculate remaining intervals and reprogram the timer at each tick interrupt. To give a feel for the overhead, the RTAI group claims that on a 233 MHz Pentium III, periodic mode supports tick rates up to 90 kHz while one-shot mode supports up to about 30 kHz.

One-shot mode is invoked by calling rt_set_one_shot_mode(). There is also a function rt_set_periodic_mode(), which happens to be the default.

The "o" run-time parameter for rt_process sets up one-shot timing mode. Take a look at the source code of rt_process down around line 85. When one-shot mode is invoked the argument to start_rt_timer() is not used. In principle, the translation from nanoseconds to counts could be different in one-shot mode. In practice, on Intel processors, it's the same. Note that we could pull the two lines,

```
period = nano2count (PERIOD);
start_rt_timer (period);
```

outside the if statement. But it was done this way to make the above distinctions clearer.

Try it out and see if there are any noticeable differences in jitter between periodic and one-shot timer modes.

Moving to Kernel Space

LXRT is a good way to get started with RTAI because you can use DDD to see what's going on. But now it's time to see what happens in Kernel Space. Take a look at rt_module.c in TaskJitter/. Go down to the init_module() function around line 83. The first thing to notice is that rt_task_init() takes more arguments than the LXRT form. In this case we have to specify a function, latency(), that implements the task as well as the size of the task's stack. On the other hand, we don't need to give the task a name. Note also that oneshot is now a module parameter.

init_latency(), the module initialization function, does basically the same initialization as rt_process.c Likewise exit_latency() does the same basic post processing as rt_process.c. Look at the function latency() starting at line 33. It first registers the mailbox name so that the check task running in User Space can find it. Then it enters almost the exact same infinite loop as rt_process. The only difference is that it doesn't test for a message from the check task. The only sensible way to stop a Kernel Space task is to remove the module.

Note the use of rt_printk() inside latency(). printk() itself is not safe to use inside an RTAI task. printk() "thinks" it has disabled interrupts but of course it has only disabled Linux interrupts. The "real" interrupts are under control of RTAI. rt_printk() manages interrupts at the RTAI level and thus is safe to use within RT tasks.

Try it out. On my system the numbers all come back zero, even while loading Netscape. Hmmm… That's enough to make you suspicious that something's wrong here. Check the code. I think you'll find that the algorithm inside the loop in latency() is identical to the loop in rt_process.c. I even tried setting samp.min and samp.max to nonzero values before the call to rt_mbx_receive() just to be sure the data structure was being written from the mailbox. It is. We've achieved genuine hard real-time performance.

RTAI /proc Files

Upon installing the rtai module, you'll find a new subdirectory under /proc, /proc/rtai. Each RTAI module creates its own file in /proc/rtai to convey its own status information. Just for kicks, take a look at /proc/rtai/scheduler both before and after installing rtai_lxrt. The scheduler file lists the real-time tasks and useful information about them. After installing rtai_lxrt we find 16 tasks in the list! Look at /proc/rtai/lxrt and you'll find that it too has created 16 tasks. If you start rt_process and look at /proc/rtai/lxrt again, you'll find two more objects in the list, the LATCAL task and LATMBX mailbox.

Real-Time FIFOs and Shared Memory

cd $(HOME/)Rtdemos/FIFO and open the file data_acq.c. This is a pseudo data acquisition application that uses an RTAI FIFO and a shared memory region to communicate with a logger process. data_acq supports multiple channels and we can control the acquisition process through a channel_t data structure defined in data_acq.h. We can, for example, set the "sample rate" and "gain" independently for each channel. In this pseudo version data_acq simply generates a sawtooth waveform on each channel at the specified rate and with a specified range.

Go to the module initialization function, init_data_acq() at line 89. The first thing it does is to create a FIFO for transferring data from the RT task to the logger. The FIFO is identified by an integer and we arbitrarily set the size to 1024 bytes. Next we create a signal handler for the FIFO. This function is a callback that is invoked whenever the User Space end of the FIFO is accessed.

It's quite possible for the data FIFO to fill up, especially if the data_acq task is started before the logger. In that case data_acq can pend on a semaphore until the logger reads something out of the FIFO thereby making room available again. The FIFO signal handler posts the semaphore when it is invoked by the data FIFO being read and if the data_acq task is pended. init_data_acq() creates the semaphore.

Next we allocate a shared memory region for the channel_t control structures. rt_shm_alloc() is the RTAI version of malloc() and kmalloc(). RTAI shared memory uses the same object name space as LXRT to allow User Space processes to access the shared memory. Finally init_data_acq() initializes and starts the data_acq task as we've seen before.

The data_acq task begins at line 45. By convention, a channel is not enabled for acquisition if its sample_period field is 0. All channels are initialized to disabled. The logger will set appropriate operational values for each channel.

In the main loop, each time the task wakes up it loops through all the channels in the channel[] array to find any that require a data sample to be generated. For each channel whose sample period has expired, it fills in a data_point_t structure, including channel number, data value, and time stamp, and sends this to the data FIFO. The function value of rtf_put() is the number of bytes written, which may be less than the number requested if the FIFO fills up. In this case data_acq pends on the semaphore.

If in fact data_acq writes less than a full data_point_t record to the FIFO, it might be possible for the logger to get out of sync by reading a partial record. It works out in this instance because the data_point_t record is 16 bytes, an integral submultiple of the FIFO size. So either a full record will be written or nothing will be written. Likewise, on the logger's end either a full record is read or the FIFO is empty. Note also that when the FIFO fills up, the current data point is simply thrown away. Is that good policy? Well, if the FIFO fills up, the task blocks and we'll probably lose data points anyway. Better that the logger be able to keep up.

Now take a look at logger.c. It starts out with an array of channel_t structures that initialize four channels. A signal handler intercepts Control-C to terminate the program gracefully. Logger writes the data it receives from data_acq to a disk file whose name is passed as a run-time parameter. Note how the FIFO is opened as an ordinary file for read-only access.

logger uses the same function as data_acq, rtai_malloc() to attach to the shared memory region. The first call to rtai_malloc() for a given region name causes it to be allocated. Subsequent calls simply make the connection. After copying the channel_t data structures to shared memory, logger just reads the data FIFO and writes to disk until it's terminated.

Remember that RT FIFOs appear in User Space as ordinary character devices. Before we can use them, we have to create inodes in the /dev directory to represent these devices. We'll need two FIFOs for our purposes, so as Superuser, execute the following two commands:

```
mknod /dev/rtf0 150 0
mknod /dev/rtf1 150 1
```

Make sure they're readable and writeable by everyone. Actually, the rtai_fifo module is set up to handle up to 64 FIFOs, if you want to create the rest of the device nodes.

Try it out. After making the targets, you'll need to insmod the following modules if they're not already installed:

```
rtai_hal.ko
rtai_ksched.ko
rtai_fifos.ko
rtai_shm.ko
data_acq.ko.
```

Of course, the .runinfo file in the FIFO/ directory takes care of this for you. Then run logger, passing it a filename as a parameter. Let it run for maybe 10 seconds to acquire some data then type Control-C to stop it. Use the following command to examine the file created by logger and verify that it really did log data.

```
od –t xI datafile > datafile.txt
```

Where *datafile* is the filename you passed to logger. od translates a binary file into a readable format (octal by default) and dumps it to stdout. –t specifies the output format, in this case hexadecimal (x) integers (I).

Suggested Exercises

- This has been a fairly brief tour of the major features of RTAI. Having gone through the data acquisition example, here are some suggested projects for exploring further.
- Create one or more simple utilities to control the data acquisition parameters for a specified channel such as sample period and range or gain. A utility to return the number of samples acquired on a given channel would also be useful.

- Create a /proc file that returns the basic channel parameters for all four channels.
- Port the thermostat example to RTAI. Provide a utility to adjust the set-point through shared memory. The current temperature could either be sent through a FIFO or also put in shared memory.
- Incorporate the parallel port hardware and device driver of Chapter 7 into the RTAI thermostat. Remember that RTAI tasks run in Kernel Space so they can directly access I/O ports. In fact, you could think of RTAI as a sort of "super device driver" model.

Resources

There's another very extensive set of examples for RTAI in the file showroom.tar.gz in the RTAI directory on the book CD-ROM. This comes from rtai.org itself and illustrates many more features of the system as well as the elaborate, involved build process.

CHAPTER 12

Posix Threads

Posix, also written POSIX, is an acronym that means Portable Operating System Interface with an X thrown in for good measure. Posix represents a collection of standards defining various aspects of a portable operating system based on UNIX. These standards are maintained jointly by the Institute of Electrical and Electronic Engineers (IEEE) and the International Standards Organization (ISO). Recently the various documents have been pulled together into a single standard in a collaborative effort between the IEEE and The Open Group (see the Resources section below).

In general, Linux conforms to Posix. The command shell, utilities and system interfaces have all been upgraded over the years to meet Posix requirements. But in the context of real-time multitasking, as in RTAI, we are specifically interested in the Posix Threads interface known as *1003.1c*.

There is another advantage to using the Pthreads API. Linux itself implements Pthreads in User Space. In effect Pthreads becomes a portable alternative to LXRT. In principle, you can develop your real-time application in User Space, using DDD/GDB for debugging, and then port it with relatively little change to Kernel Space RTAI. Unfortunately, as we'll see later, it's not quite that simple.

The header file that prototypes the Pthreads API is `pthread.h` and resides in the usual directory for library header files, `/usr/include`.

Threads

Fundamentally a thread is the same thing we were calling a task in Chapter 9. It is an independent thread of execution embodied in a function. The thread has its own stack.

```
int pthread_create (pthread_t *thread, pthread_attr_t *attr, void
    *(* start_ routine) (void *), void *arg);
void pthread_exit (void *retval);
int pthread_join (pthread_t thread, void **thread_return);
pthread_t pthread_self (void);
int sched_yield (void);
```

The mechanism for creating and managing a thread is analogous to creating and managing the tasks. A thread is created by calling **pthread_create()** with the following arguments:

- *pthread_t* – A *thread object* that represents or identifies the thread. **pthread_create()** initializes this as necessary.
- Pointer to a thread *attribute* object. Often it is NULL. More on this later.
- Pointer to the start routine. The start routine takes a single pointer to void argument and returns a pointer to void.
- Argument to be passed to the start routine when it is called.

A thread may terminate by calling **pthread_exit()** or simply returning from its start function. The argument to **pthread_exit()** is the start function's return value.

In much the same way that a parent process can wait for a child to complete by calling **waitpid()**, a thread can wait for another thread to complete by calling **pthread_join()**. The arguments to **pthread_join()** are the thread object of the thread to wait on and a place to store the thread's return value. The calling thread is blocked until the target thread terminates. RTAI Pthreads doesn't implement **pthread_join()** as the tasking model doesn't support the notion of joining .

A thread can determine its own ID by calling **pthread_self()**. Finally, a thread can voluntarily yield the processor by calling **sched_yield()**.

Note that most of the functions above return an int value. This reflects the Pthreads approach to error handling. Rather than reporting errors in the global variable **errno**, Pthreads functions report errors through their return value. Appendix B gives a more complete description of the Pthreads API including a list of all error codes.

Thread Attributes

Posix provides an open-ended mechanism for extending the API through the use of *attribute objects*. For each type of Pthreads object there is a corresponding attribute object. This attribute object is effectively an extended argument list to the related object create or initialize function. A pointer to an attribute object is always the second argument to a create function. If this argument is NULL the create function uses appropriate default values. This also has the effect of keeping the create functions relatively simple by leaving out a lot of arguments that normally take default values.

An important philosophical point is that all Pthreads objects are considered to be "opaque." Most of them anyway. We'll see an exception shortly. This means that you never directly access members of the object itself. All access is through API functions that get and set the member arguments of the object. This allows new arguments to be added to a Pthreads object type by simply defining a corresponding pair of get and set functions for the API. In simple implementations the get and set functions may be nothing more than a pair of macros that access the corresponding member of the attribute data structure.

```
int pthread_attr_init (pthread_attr_t *attr);
int pthread_attr_destroy (pthread_attr_t *attr);
int pthread_attr_getdetachstate (pthread_attr_t *attr, int *detachstate);
int pthread_attr_setdetachstate  (pthread_attr_t *attr, int detachstate);

Scheduling Policy Attributes
int pthread_attr_setschedparam (pthread_attr_t *attr, const struct
sched_param *param);
int pthread_attr_getschedparam (const pthread_attr_t *attr, struct
sched_param *param);
int pthread_attr_setschedpolicy (pthread_attr_t *attr, int policy);
int pthread_attr_getschedpolicy (const pthread_attr_t *attr, int *policy);
int pthread_attr_setinheritsched (pthread_attr_t *attr, int inherit);
int pthread_attr_getinheritsched (const pthread_attr_t *attr, int *inherit
```

Before an attribute object can be used, it must be initialized. Then any of the attributes defined for that object may be set or retrieved with the appropriate functions. This must be done before the attribute object is used in a call to pthread_create(). If necessary, an attribute object can also be "destroyed." Note that a single attribute object can be used in the creation of multiple threads.

The only required attribute for thread objects is the "detach state." This determines whether or not a thread can be joined when it terminates. The default detach state is PTHREAD_CREATE_JOINABLE meaning that the thread can be joined on termination. The alternative is PTHREAD_CREATE_DETACHED, which means the thread can't be joined.

Joining is only useful if you really need the thread's return value. Otherwise it's better to create the thread detached. The resources of a joinable thread can't be recovered until another thread joins it whereas a detached thread's resources can be recovered as soon as it terminates. In RTAI Pthreads the only valid value for detach state is PTHREAD_CREATE_DETACHED and pthread_attr_setdetachstate() doesn't really do anything.

There are also a number of optional scheduling policy attributes that RTAI Pthreads implements. See Appendix B for more details on these.

Synchronization—Mutexes

Pthreads uses the mutex as its primary synchronization mechanism.

```
pthread_mutex_t mutex = PTHREAD_MUTEX_INITIALIZER;

int pthread_mutex_init (pthread_mutex_t *mutex, const
    pthread_mutexattr_t *mutex_attr);
int pthread_mutex_destroy (pthread_mutex_t *mutex);

int pthread_mutex_lock (pthread_mutex_t *mutex);
int pthread_mutex_unlock (pthread_mutex_t *mutex);
int pthread_mutex_trylock (pthread_mutex_t *mutex);
```

The Pthreads mutex API follows much the same pattern as the thread API. There is a pair of functions to initialize and destroy mutex objects and a set of functions to act on the mutex objects. The listing also shows an alternate way to initialize statically allocated mutex objects. PTHREAD_MUTEX_INITIALIZER provides the same default values as pthread_mutex_init().

Two operations may be performed on a mutex: *lock* and *unlock*. The lock operation causes the calling thread to block if the mutex is not available. There's another function called *trylock* that allows you to test the state of a mutex without blocking. If the mutex is available *trylock* returns success and locks the mutex. If the mutex is not available it returns EBUSY

Mutex Attributes

```
int pthread_mutexattr_init (pthread_mutexattr_t *attr);
int pthread_mutexattr_destroy (pthread_mutexattr_t *attr);

int pthread_mutexattr_getkind_np (pthread_mutexattr_t *attr, int
*kind);
int pthread_mutexattr_setkind_np (pthread_mutexattr_t *attr, int
kind);

kind =     PTHREAD_MUTEX_FAST_NP
           PTHREAD_MUTEX_RECURSIVE_NP
           PTHREAD_MUTEX_ERRORCHECK_NP

int pthread_mutexattr_getprioceiling (const pthread_mutexattr_t
    *mutex_attr, int *prioceiling);
int pthread_mutexattr_setprioceiling (pthread_mutexattr_t *mutex_
    attr, int prioceiling);
int pthread_mutexattr_getprotocol (const pthread_mutexattr_t
    *mutex_attr, int *protocol);
int pthread_mutexattr_setprotocol (pthread_mutexattr_t *mutex_attr,
    int protocol);

protocol =    PTHREAD_PRIO_NONE
              PTHREAD_PRIO_INHERIT
              PTHREAD_PRIO_PROTECT
```

Mutex attributes follow the same basic pattern as thread attributes. There is a pair of functions to create and destroy a mutex attribute object. The only mutex attribute in RTAI Pthreads is a Linux-specific nonportable extension called *kind*. The Pthreads standard explicitly allows nonportable extensions. The only requirement is that any symbol that is nonportable have "_np" appended to its name as shown here.

What happens if a thread should attempt to lock a mutex that it has already locked? Normally the thread would simply hang up. Linux offers a way out of this dilemma. The "kind" attribute alters the behavior of a mutex when a thread attempts to lock a mutex that it has already locked. The possible values for kind are:

Fast. This is the default type. If a thread attempts to lock a mutex it already holds it is blocked and thus effectively deadlocked. The fast mutex does no consistency or sanity checking and thus it is, usually, the fastest implementation.

Recursive. A recursive mutex allows a thread to successfully lock a mutex multiple times. It counts the number of times the mutex was locked and requires the same number of calls to the unlock function before the mutex goes to the unlocked state.

Error checking. If a thread attempts to recursively lock an error checking mutex, the lock function returns immediately with the error code EDEADLK. Furthermore, the unlock function returns an error if it is called by a thread other than the current owner of the mutex.

Note the "`_np`" in the constant names.

Optionally a Pthreads mutex can implement the priority inheritance or priority ceiling protocols to avoid priority inversion as discussed in Chapter 9. The mutex attribute *protocol* can be set to "none," "priority inheritance," or "priority ceiling." The *prioceiling* attribute sets the value for the priority ceiling. The protocol and prioceiling attributes are not available in RTAI Pthreads.

Communication—Condition Variables

There are many situations where one thread needs to notify another thread about a change in status to a shared resource protected by a mutex. Consider the situation in Figure 12-1 where two threads share access to a queue. Thread 1 reads the queue and Thread 2 writes it. Clearly each thread requires exclusive access to the queue and so we protect it with a mutex.

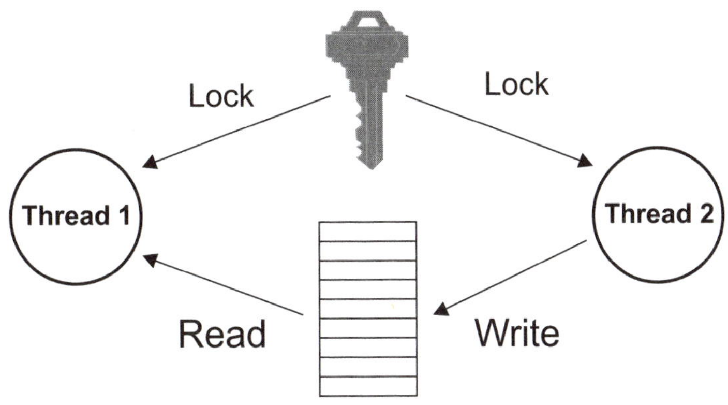

Figure 12-1: Condition Variable

Thread 1 will lock the mutex and then see if queue has any data. If it does, Thread 1 reads the data and unlocks the mutex. However, if the queue is empty, Thread 1 needs to block somewhere until Thread 2 writes some data. Thread 1 must unlock the mutex before blocking or else Thread 2 would not be able to write. But there's a gap between the time Thread 1 unlocks the mutex and blocks. During that time, Thread 2 may execute and not recognize that anyone is blocking on the queue.

The condition variable solves this problem by waiting (blocking) with the mutex locked. Internally, the conditional wait function unlocks the mutex allowing Thread 2 to proceed. When the condition wait returns, the mutex is again locked.

```
pthread_cond_t cond = PTHREAD_COND_INITIALIZER;

int pthread_cond_init (pthread_cond_t *cond, const pthread_condattr_
    t *cond_attr);
int pthread_cond_destroy (pthread_cond_t *cond);

int pthread_cond_wait (pthread_cond_t *cond, pthread_mutex_t *mutex);
int pthread_cond_timedwait (pthread_cond_t *cond, pthread_mutex_t
    *mutex, const struct time-spec *abstime);
int pthread_cond_signal (pthread_cond_t *cond);
int pthread_cond_broadcast (pthread_cond_t *cond);
```

The basic operations on a condition variable are *signal* and *wait*. Signal wakes up one of the threads waiting on the condition. The order in which threads wake up is a function of scheduling policy. A thread may also execute a *timed wait* such that if the specified time interval expires before the condition is signaled, the wait returns with an error. A thread may also *broadcast* a condition. This wakes up all threads waiting on the condition.

Condition Variable Attributes

Pthreads does not define any required attributes for condition variables although there is at least one optional attribute. RTAI Pthreads provides the functions to initialize and destroy a condition variable attribute object but does not implement the optional attribute.

Pthreads in User Space

Until kernel version 2.6, the most prevalent threads implementation was LinuxThreads. It has been around since about 1996 and by the time development began on the 2.5 kernel it was generally agreed that a new approach was needed to address the limitations in LinuxThreads. Among these limitations, the kernel represents each thread as a separate process, giving it a unique process ID, even though many threads exist within one process entity. This causes compatibility problems with other thread implementations. There's a hard coded limit of 8192 threads per process, and while this may seem like a lot, there are some problems that can benefit from running thousands of threads.

The result of this new development effort is the Native Posix Threading Library or NPTL, which is now the standard threading implementation in 2.6 series kernels. It too is a one-to-one model, but takes advantage of improvements in the kernel that were specifically intended to overcome the limitations in LinuxThreads. The `clone()` call was extended to optimize thread creation. There's no longer a limit on the number of threads per process, and the new fixed time scheduler can handle thousands of threads without excessive overhead. A new synchronization mechanism, the Fast Mutex or "futex," handles the noncontention case without a kernel call.

In tests on an IA-32, NPTL is reportedly able to start 100,000 threads in two seconds. By comparison, this test under a kernel without NPTL would have taken around 15 minutes.

As an exercise in using Pthreads in User Space we'll recast the data_acq example from the last chapter. This is again a 4-channel simulated data acquisition application where each channel generates a simple sawtooth of a specified amplitude. The acquired data is displayed on the screen rather than written to a file. The primary difference is that rather than have a single task manage all four channels, each channel now gets its own thread. The motivation behind this approach is that a thread that only manages one channel is probably simpler than a thread that manages "n" channels.

Each of the "channel" threads writes its data into a common data structure protected by a condition variable. A single "display" thread reads the structure and writes the data to the screen. Finally, a "command" thread accepts simple operator commands from the keyboard and updates the channels accordingly. The system architecture is graphically illustrated in Figure 12-2.

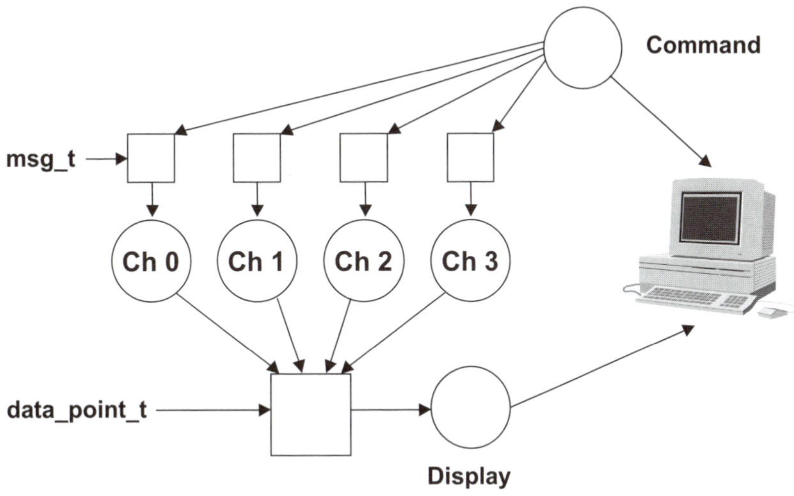

Figure 12-2: data_acq in User Space

cd /$HOME/Rtdemos/Posix and open data_acq.c. Near the top at line 21 is an array of channel_t data structures. Looking at data_acq.h, you'll see that many of the fields in channel_t have been removed. That's because they can now be local variables inside the data_acq() thread. Other fields have been added to support Pthreads. Note that the data_point_t typedef includes an #ifdef to specifiy an alternate definition of timestamp. It turns out that we need to deal with time a little differently in Kernel Space.

Next is a data_point_t structure for communicating data from the channel threads to the display thread. This structure requires a mutex and a condition variable to protect it.

Before looking at the data_acq() thread, let's move down to main() at line 166. It starts by initializing the four data acquisition channels using the channel[] array. For each channel it creates a mutex, a condition variable, and a thread. The argument passed to the thread is the channel's channel_t structure.

Next we initialize the display. We're using the Curses library as we did in devices.c back in Chapter 6. Then we create the Command() thread that will monitor the keyboard for operator input. Finally, main() calls Display(), which becomes the main thread that waits for data from the channel threads and displays it. We could have just left the display loop in main() but when it's time to move to Kernel Space it will be useful to have a separate function.

Move up to Display() at line 146. Here then is our first example of using mutexes and condition variables. The display thread locks the mutex associated with the display data structure (m_display) and waits on the corresponding condition variable (c_display). pthread_cond_wait() won't return until someone else, one of the channel threads, signals that it has put something in the display structure. Upon waking up, the display thread copies the data to a local variable before unlocking the mutex.

Now let's go back and look at data_acq() starting at line 26. We first copy a number of fields from the channel_t structure passed as an argument. This is as much for readability as anything else. We maintain the same convention that a sample period of zero means the channel is disabled. data_acq() is now ready to go to sleep and here's where it gets interesting. There are two different circumstances in which data_acq() should wake up:

1. It's sampling and the sample period has expired
2. It received a message from Command() to alter one of its sampling parameters

data_acq() receives messages from Command() through a buffer contained in its channel_t structure. Since this buffer is accessed by two threads, it must be protected by a mutex. We lock the associated mutex in the channel_t structure before accessing the message buffer. But there's no point in trying to read the message buffer until the Command() thread sends us something. So we'll wait on the condition variable in channel_t and let Command() signal us when something changes.

If the channel is not sampling (**sample_period** == 0), then we can just call **pthread_cond_wait()** until the operator decides to turn this channel on. If the channel is sampling, we could call some "sleep" function to just delay for the sample period but at the same time, we also have to respond to operator commands. Here's where **pthread_cond_timedwait()** is useful. We can call **pthread_cond_timedwait()** with the sample period as the timeout value. When it returns, the status value will indicate whether the call timed out (**status** == ETIMEOUT) or somebody signaled us (**status** == 0).

Upon waking up from **pthread_cond_timedwait()**, we have to compute the next wakeup time. That's because timed wait takes absolute time as its timeout argument[1]. So we add **sample_period**, converted to nanoseconds, to the timeout argument.

If status indicates a timeout, we're not interested in the message buffer so we can immediately unlock the mutex. Now we need exclusive access to the display data structure so we lock its mutex. Then we copy the relevant information to the display structure, signal its condition variable and unlock the mutex.

If **status** is zero, **Command()** has put something into the message buffer. In this case we leave the mutex locked while we deal with the message. The major complication here is that if the message changes the sample period, we have to compute a new wakeup time.

Let's move on to the **Command()** thread at line 108. The **getstr()** function blocks the thread until the operator types <Enter>. Upon returning, we parse the command string. The command syntax is similar to what you did in Chapter 6 adding programmable parameters to the thermostat. There are four valid commands:

- "c" – *channel*. The **Command()** thread handles this internally. Subsequent commands are directed to this channel until the channel value is changed.
- "q" – *quit*. Stop everything. The call to **exit()** cancels all threads. This is probably not the cleanest way to stop the system. The clean solution would be for **Command()** to notify the other threads that it's time to go away and have them commit suicide.

[1] I find this somewhat "klutzy." Other systems I'm familiar with tend to treat timeouts as intervals rather than absolute time. You just call the function again with the same timeout interval. But more than klutzy, this is just the sort of thing we don't need in embedded systems. Y2K was not the disaster that the popular press predicted precisely because the vast majority of embedded systems have no concept of absolute time. Fortunately, Unix absolute time won't overflow until Feb. 7, 2106, but if you happen to treat time as a signed value, it overflows on Jan. 19, 2038. Hmmm…

- "r" – *range*. The maximum value for data generated by this channel.
- "s" – *sample period*. Set the sample period for the current channel in milliseconds.

In the interest of expediency, Command() does no error or sanity checking. Except for "c" and "q" (case sensitive by the way), all perceived command tokens are passed on to the current channel. There's no check that a channel value is in range.

Keyboard input is not echoed. This would require sharing the display between two threads so that Command() could move the cursor to a command line field and then restore it as each character is typed. This in turn requires that the display be protected by a mutex[2].

Making data_acq

Have a look at the Makefile. Ignore the commented section, we'll come back to that later. The logger target also isn't relevant yet. Two libraries have been added to the compile command for data_acq with the "–l" option. The names are fairly obvious. The pthread library contains the Pthreads functions and curses is the curses library.

There's also a compile time symbol, _REENTRANT. The original Linux library routines assumed that only a single thread of execution would be running in a process. An example of this assumption is the global variable errno. That works fine if there's only a single execution thread. But when multiple threads are executing "simultaneously," it would be quite possible for a function called by one thread to set errno only to have the value changed by a function called by a second thread before the first thread read the value intended for it. That's why Pthreads functions return their error codes as the function value rather than using errno.

But it goes deeper than that. Functions that could be called simultaneously by more than one thread must be *reentrant*. A function is reentrant if it doesn't use any statically allocated resources such as global memory. C functions tend to be naturally reentrant provided they only reference local variables because these variables are allocated on the stack. Each thread gets its own copy of the local variables. But some library functions, the printf() family for example, use statically allocated buffers. The purpose of the _REENTRANT flag then is to alert the compiler to substitute reentrant versions of these functions that are "thread safe."

[2] Another of those proverbial "exercises for the reader." Have a go at it.

So try it out. make data_acq and run it. Enter the command "c 0 s 500" to start channel 0 sampling. Then enter similar commands for the other channels.

Debugging Multithreaded Programs

The nice thing about running Pthreads in User Space is that you can debug your program with DDD/GDB. Multithreading does introduce some complications to the debugging process but, fortunately, GDB has facilities for dealing with those. Run data_acq under DDD and set the initial breakpoint at the line:

```
channel[i].number = i;
```

near the beginning of main(). But before running the program, execute View->Execution Window to bring up a separate window for the program to write its output. Curses output doesn't work very well in the gdb console window.

Now run the program and when it stops at the breakpoint you'll see the following messages in the gdb console window:

```
[New Thread 1024 (runnable)]
[Switching to Thread 1024 (runnable)]
```

GDB has recognized that it's running a multithreaded environment. Let the program continue. When it stops at that breakpoint again, you'll see two more "New Thread" messages with different numbers. One of those is the channel thread created by the "for" loop, the other is created by Pthreads for its own use. Let the program continue again and another New Thread message will appear.

Delete the breakpoint and let the program continue to the while (1) statement at the top of the display loop. At this point all the threads have been created. Execute Status->Threads. You'll see a list of all the threads and their current execution point. The currently executing thread is highlighted. Note that four of the threads, the channels, are all at the same execution point.

Execute Status->Backtrace to bring up the call stack for the current thread. This shows that we're in main() and that main() was called from __libc_start_main(). Select one of the threads that's at sigsuspend.c:48. The Backtrace display immediately changes to show the call stack of that thread. So you can make any thread the "active thread" and see its status.

If you set a breakpoint in the data_acq() function, the program will stop when any of the four threads executing data_acq() reaches that point. You can then examine the local variables for that thread. Note however that only the local variables for the currently executing thread are visible.

Moving to RTAI Kernel Space

Oddly, the RTAI Pthreads implementation seems to have taken a step backwards in terms of portability. When I wrote the first edition of this book, RTA 24.1.9 included a Kernel Space Pthreads module using the standard API calls. Granted it was incomplete, but for the implemented APIs, you could move fairly easily from User Space Pthreads to Kernel Space RTAI Pthreads.

The Pthreads implementation in RTAI 3.1 is a set of in-line function wrappers around the native RTAI APIs. These are implemented in the header file rtai_posix.h in rtai_core/include/. The odd thing is that all the wrapper functions have "_rt" appended to the function name so they are no longer portable. A search of the RTAI mail list archive reveals very little about the motivation for this beyond a comment by Paolo Mantegazza that it was done to "avoid incompatibilities."

The Posix/ directory contains an alternate version of data_acq.c called data_acq_rt.c properly reconfigured to run in RTAI kernel space. Most of the functionality of main() becomes the module init function init_data_acq(). And of course we need a cleanup function. In addition to creating the Command() thread we now have to explicitly create the Display() thread. printf() changes to printk().

The major complication of course is that we can't really do any screen I/O from Kernel Space. Among other things, we don't have access to the curses library from Kernel Space. So the real functionality of Display() and Command() needs to be split between Kernel Space and User Space. We use RTAI FIFOs to communicate between the User Space threads and their "buddies" running in Kernel Space. The curses initialization calls in init_data_acq() are replaced with two rtf_create() calls. data_acq.h contains #defines for the two FIFOs.

Kernel Space Display() simply copies the data points it receives to a FIFO. The getstr() call in Command() is just replaced by a rtf_get() call.

The function ftime() is not available in Kernel Space so we'll need an alternative for reading time. RTAI Pthreads defines:

```
void clock_gettime_rt (int clock_id, struct timespec *current_time);
```

which is the kernel equivalent of the Posix function clock_gettime(). In this case the clock_id argument is not used. struct timespec is defined in time.h and has two long int fields:

tv_sec – seconds since midnight Jan. 1, 1970

tv_nsec – nanoseconds since tv_sec

This is almost identical to the timeb structure that data_acq.c uses.

logger.c implements the User Space "buddies" for Command() and Display().

Now open up Makefile. The section that is commented out with '#' builds the kernel module data_acq_rt.o. Delete the '#' characters and run make again.

data_acq_rt.o requires the following modules:

rtai_hal.ko

rtai_ksched.ko

rtai_sem.ko

rtai_mq.ko

rtai_fifos.ko

data_acq_rt.ko

Then run ./logger. If everything is working correctly you should be able to enter "c 0 s 500" and see channel 0 start "acquiring data". If not, then it's time to add rt_printk() statements at strategic points in data_acq_rt.c to see what's going on.

Message Queues

Message queues are an element of the overall Posix standard defined as part of the optional real-time extensions. The real-time extensions are not part of standard Linux, at least not yet, but Posix message queues have been implemented in RTAI. So it seems reasonable to include them in this chapter as an element of RTAI's "Posixness". Message queues are implemented as a separate module, rtai_mq.ko, with a corresponding header file rtai_mq.h.

Perhaps because it's a real module and not a set of in-line function wrappers, the RTAI message queue API doesn't append the "_rt" to the function names.

Message queues are very similar to RTAI FIFOs except that they are full duplex and they deal in discrete messages rather than continuous streams of bytes. Unfortunately, they only work between tasks in Kernel Space.

```
mqd_t mq_open (char *mq_name, int oflags, mode_t permissions,
    struct mq_attr *mq_attr);
int mq_close (mqd_t mq);
int mq_unlink (char *mq_name);

size_t mq_receive (mqd_t mq, char *msg_buffer, size_t buflen, unsigned
    int *msgprio);
int mq_send (mqd_t mq, const char *msg, size_t msglen, unsigned
    int msgprio);
int mq_notify (mqd_t mq, const struct sigevent *notification);

int mq_getattr (mqd_t mq, struct mq_attr *attrbuf);
int mq_setattr (mqd_t mq, const struct mq_attr *new_attrs, struct
mq_attr *old_attrs);
```

The message queue API is relatively simple and straightforward. A message queue is opened and closed much like a file. When opening a message queue you specify access mode and permissions. The open function also has a Pthreads flavor by including an attributes structure. Closing a message queue does not destroy it but merely destroys the link to it. The queue and any messages in it remain available to other links and other threads that may open links to it.

A queue is destroyed, and its resources freed, by *unlinking* it. However, if there are any open links to the queue when mq_unlink() is called, the queue is marked for destruction when the last link is closed. Once a queue has been marked for destruction no further links may be opened to it.

The operations on a queue are just *receive* and *send*. Both operations are blocking unless the O_NONBLOCK flag is set when the link is opened. Messages are placed in the queue in priority order. That is, a higher priority will be inserted ahead of a lower priority message already there. Messages of equal priority are queued in FIFO order.

mq_notify() allows a thread to specify an asynchronous callback to be called when a message arrives in the queue. The struct sigevent contains, among other things, a pointer to the function to be called. RTAI messages queues implements mq_notify() for completeness but does not in fact do asynchronous notification.

The message queue API takes a different approach to attributes than Pthreads. Rather than having a pair of get and set functions for each attribute, the attribute structure fields are made explicitly visible as shown in Listing 12-1. The pair of get and set attributes functions takes a pointer to this structure. When creating a queue you can specify the maximum size of a message and the maximum number of messages that the queue hold. Subsequently the only thing that can be changed by mq_set_attr() is the blocking behavior specified in mq_flags.

```
In file rtai_mq.h

struct mq_attr {
   long mq_maxmsg;        //Maximum number of messages in queue
   long mq_msgsize;       //Maximum size of a message (in bytes)
   long mq_flags;         //Blocking/Non-blocking behaviour specifier
                          // not used in mq_open only relevant for
                          // mq_getattrs and mq_setattrs
   long mq_curmsgs;       //Number of messages currently in queue
};
```

Listing 12-1

The RTAI message queue implementation is another good example of Pthreads programming. Have a look at /usr/src/rtai/rtai-core/ipc/mq/mq.c. Also, as Linux code goes, it is surprisingly well commented.

Suggestions for Further Exploration

If you've managed to get data_acq_rt working and had a look at the message queue implementation, mq.c, you've gained a pretty good understanding of the Posix features of RTAI. Consider hacking RTAI itself. Pthreads in RTAI 3.1 leaves out a number of useful features. For example:

- *RTAI doesn't implement mutex attributes*. What would it take to implement the protocol attribute for mutexes? The basic issue here is changing a task's priority. RTAI does provide a function for this but it's not publicly documented. Take a look at the code in /usr/src/rtai/rtai-core/sched/rtai.
- *RTAI doesn't implement the "detach state" thread attribute*. All threads are joinable. What would it take to create a detachable thread?

- It would really be nice if message queues could be used transparently between Kernel Space and User Space like RTAI FIFOs. `/usr/src/rtai/rtai-core/ipc/fifos/fifos.c` provides a good starting point.

Of course, before embarking on any serious update to the RTAI code, you should check the current status. Someone may have already started working on the same project. There are two very active mail lists for RTAI, one for users and one for developers. You can subscribe by going to *www.rtai.org*.

Resources

The Open Group has made available free for private use the entire Posix (now called *Single Unix*) specification. Go to:

www.unix.org/online.html

You'll be asked to register. Don't worry, it's painless and free. Once you register you can read the specification online or download the entire set of html files for local access.

This chapter has been little more than a brief introduction to Posix threads. There's a lot more to it, even in the RTAI implementation. An excellent, more thorough treatment is found in,

Butenhof, David R., *Programming with POSIX Threads*, Addison-Wesley, 1997.

CHAPTER 13

Cutting It Down to Size

"A designer knows he has achieved perfection not when there is nothing left to add, but when there is nothing left to take away."
— Antoine de Saint-Exupery

Often, the biggest problem in an embedded environment is the lack of resources, specifically memory and storage space. As you've no doubt observed, either in the course of reading this book, or from other experience, Linux is *big*! The kernel itself is often well over a megabyte, and then there are all the attendant utility programs that have to go somewhere. A minimum Red Hat or Debian installation comes out to about 40 Mbytes of file storage.

In this chapter, we'll look at some strategies for substantially reducing the overall "footprint" of Linux to make it fit in limited resource embedded devices.

BusyBox

Even if your embedded device is "headless," that is it has no screen and/or keyboard in the usual sense for user interaction, you still need a minimal set of command line utilities. You'll no doubt need **mount**, **ifconfig**, and probably several others to get the system up and running. Remember that every shell command line utility is a separate program with its own executable file.

The idea behind BusyBox© is brilliantly simple. Rather than have every utility be a separate program with its attendant overhead, why not simply write one program that implements *all* the utilities? Well, perhaps not all, but a very large number of the most common utilities. Most utilities require the same set of "helper" functionality. Rather than duplicating these functions in dozens of files, BusyBox implements them exactly once.

In many cases, the BusyBox version of the utilities omits some of the more obscure options and switches. Think of them as "lite" versions. Nevertheless, the most useful options are preserved.

The BusyBox project began in 1996 with the goal of putting a complete Linux system on a single floppy disk that could serve as a rescue disk or an installer for the Debian Linux distribution. A *rescue disk* is used to repair a Linux system that has become unbootable. This means the rescue disk must be able to boot the system and mount the hard disk file systems, and it must provide sufficient command-line tools to bring the hard disk root file system back to a bootable state.

Subsequently, embedded developers figured out that this was an obvious way to reduce the Linux footprint in resource-constrained embedded environments. So the project grew well beyond its Debian roots and today BusyBox is a part of almost every commercial embedded Linux offering, and is found in such diverse projects as the Kerbango Internet Radio and the IBM Wristwatch that runs Linux.

BusyBox has been called the *Swiss army knife* of embedded Linux because, like the knife, it's an all-purpose tool. Technically, the developers refer to it as a "multicall binary," meaning that the program is invoked in multiple ways to execute different commands. This is done with symbolic links.

BusyBox is usually installed in the /bin directory where most user-level command line utilities reside. Then, for example, in /bin we create the following symbolic link:

```
ln -s busybox ls
```

Now when you invoke ls from a shell, what really gets executed is BusyBox. The program figures out what it's supposed to do by looking at the first argument to main(), which is the name that invoked it, that is—ls. A similar link is created for every command that BusyBox implements. Table 13.1 is the full list of commands available in BusyBox.

BusyBox is highly modular and configurable. While it is capable of implementing well over 100 shell commands, by no means are you required to have all of them in your system any more than Bluecat requires you to include all of the Linux utilities. The configuration process lets you choose exactly which commands will be included in your system.

Table 13-1: BusyBox Commands

Command	***Function***
ar	*file archiver*
basename	*delete prefix and suffix from pathname*
cat	*concatenate and show files*
chgrp	*change group ID of files*
chmod	*change file access permissions*
chown	*change owner of files*
chroot	*run command or shell with specified root directory*
chvt	*change foreground virtual terminal*
clear	*clear terminal screen*
cp	*copy files*
cut	*cut out fields from file*
date	*display current date*
dc	*arbitrary precision desk calculator*
dd	*convert and copy file*
deallocvt	*terminate unused virtual terminals*
df	*report free block count*
dirname	*delete end of pathname*
dmesg	*display or clear kernel ring buffer*
du	*summarize disk usage*
dumpkmap	*prints out a binary keyboard translation table*
dutmp	*dump utmp file*
echo	*echo arguments*
false	*return unsuccessful exit status*
fbset	*sow and modify frame buffer device settings*
fdflush	*force detection of floppy disk change*
find	*find files*
free	*display memory use*
freeramdisk	*free all memory used by a ramdisk*
fsck.minix	*check and repair Minix filesystem*
grep	*search file for pattern*

Command	Function
gunzip	*decompress files*
gzip	*compress files*
halt	*stop the system*
head	*display beginning of file*
hostid	*display a unique host identification*
hostname	*set or display host name*
id	*show user and group IDs and names*
init	*process control initialization*
insmod	*install loadable kernel module*
kill	*terminate or send a signal to processes*
killall	*kill processes by name*
length	*display length of a string*
linuxrc	*set up RAM disk root*
ln	*make links to files*
loadacm	*load an acm from standard input*
loadfont	*load a console font*
loadkmap	*load a binary keyboard translation table*
logger	*add entry to system log*
logname	*print login name from /etc/utmp*
ls	*list contents of directories*
lsmod	*list loaded modules*
makedevs	*create device files*
md5sum	*generate or check MD5 message digests*
mkdir	*create specified directories*
mkfifo	*make FIFOs (named pipes)*
mkfs.minix	*make Minix filesystem*
mknod	*make special files*
mkswapnn	*set up a swap area*
mktemp	*make temporary filenames*
more	*display file by screenfull or line*
mount	*mount a file system*
mt	*magnetic tape control*

Command	**Function**
mv	*move or rename files*
nc	*opena pipe to an IP port*
nslookup	*query internet name servers*
ping	*test network response*
poweroff	*stop system*
printf	*format and display data*
ps	*report process status*
pwd	*print working directory*
reboot	*stop system*
rm	*remove files and directories*
rmdir	*remove empty directories*
rmmod	*unload loadable modules*
sed	*stream editor*
setkeycodes	*load kernel scancode-to-keycode*
sh	*shell*
sleep	*suspend execution for specified duration*
sort	*sort/merge files*
swapoff	*disable paging and swapping*
swapon	*enable paging and swapping*
sync	*write unwritten memory info to disk*
syslogd	*linux system logging utilities*
tail	*output last part of file*
tar	*tape and file archiver*
tee	*copy stdin to stdout and files*
telnet	*connect to remote*
test	*check file types and compare values*
touch	*update file access/modification times*
tr	*translate characters*
true	*return successful exit status*
tty	*display terminal name*
umount	*unmount file systems*
uname	*display system name*

Command	Function
uniq	report repeated lines
update	flush file system buffers
uptime	display system uptime
usleep	suspend execution for specified microsecond duration
uudecode	decode ASCII representation of file
uuencode	encode ASCII representation of file
wc	count lines, words and characters
which	display pathname of command
whoami	display effective username
yes	repeat string indefinitely
zcat	display compressed file

Installing and Configuring BusyBox

Let's add BusyBox to our Bluecat target. BusyBox is available on the book CD in the root directory as busybox-1.00.tar.bz2. Untar this into bluecat/demo.x86/shell. You now have a subdirectory named busybox-1.00/. Now before doing anything in this directory, you'll have to log in as root user, or invoke the KDE File Manager in Super User mode to change the user and group settings for this directory and all directories and files under it to your user name[1].

BusyBox supports make menuconfig for configuration. The top configuration menu is shown in Figure 13-1. There are many categories of configuration options. The first three categories are general in nature, pertaining to BusyBox as a whole. There's one build option we'll want to turn on: "Build BusyBox as a static binary (no shared libs)." We don't have Bluecat configured to put shared libraries in the root filesystem, so we have to build BusyBox with the libraries static linked to it. The remaining categories are the individual commands that can be built into the package. Figure 13-2 shows the first part of the core utilities section.

[1] Once again, Linux file permissions bites you.

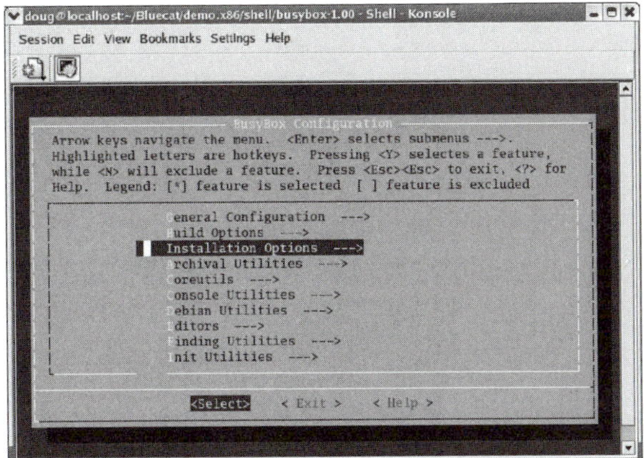

Figure 13-1: BusyBox Configuration

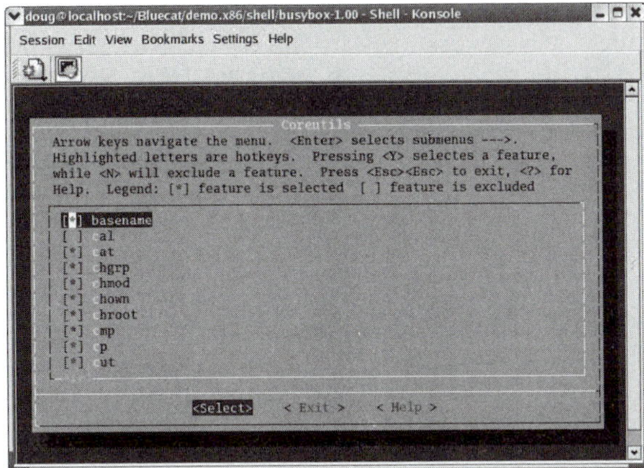

Figure 13-2: Core utilities Configuration

Browse through the various categories to get a better feel for what's there. You'll find that help is generally readable and useful. There's not a whole lot that's absolutely required for our Bluecat shell kernel. The `shell.spec` file shows what commands are copied to the root filesystem. At a minimum, we'll need to include those in BusyBox, but feel free to include others that you think might be useful. For example, ping is a useful network utility to have available in case networking doesn't come up as expected. Also, feel free to remove whatever you think doesn't need to be in the root filesystem. A lot of commands are enabled by default that we probably don't need in an embedded environment.

Note that if you select the mount command, you also need to select "Support mounting NFS file systems."

When you're finished, exit from menuconfig and execute make. This will take a while. Following the make step you'll find two new files in your busybox-1.00/ directory: the busybox executable, and a text file called busybox.links. The latter contains the necessary information for creating symbolic links for all the commands you turned on in BusyBox. Execute make install. Since we haven't specified a target installation directory, make install creates a new directory in busybox-1.00/ called _install/ with subdirectories bin/, sbin/ and usr/. bin/ has the BusyBox executable along with symbolic links for all the commands that are normally found in /bin. Likewise sbin/ has links for all the commands that normally reside in /sbin.

Of course, this isn't really what we want to do. It's simply an illustration of BusyBox's default build behavior. What we want to do is replace the command executables in our Bluecat root filesystem with BusyBox and the appropriate links. Save a copy of shell.spec as, say, shell.spec.old so you can go back to it if you have to. Now edit shell.spec. Down near the bottom of the file is a section that starts:

```
lcd ($BLUECAT_PREFIX)/bin
```

We'll replace that with the text in Listing 13-1. This copies the BusyBox executable to /bin in the target root filesystem and creates the necessary links in the appropriate directories, /bin and /sbin.

TinyLogin

This is another tool that reduces the Linux footprint by combining a number of utilities into one executable. In this case, the utilities are those that relate to the Linux login process. By combining those into one executable we can cut down even more on the Linux footprint.

TinyLogin incorporates the following commands:

- *adduser* – add new user accounts.
- *addgroup* – add new group accounts.
- *deluser* – delete users from the /etc/passwd and /etc/shadow files.
- *delgroup* – delete groups from /etc/group.
- *passwd* – change a user's password.
- *getty* – manage terminals and start the login process.

- *login* – begin a login session.
- *su* – change current user ID (substitute user, normally used to become root for a while).
- *sulogin* – single-user login (runlevel 1).
- *vlock* – lock a virtual console until a password is entered.

The last two functions are normally disabled but can be turned on at configuration time.

```
# protocols added for gdbserver
cp ./local/fstab ./local/inittab ./local/protocols /etc
cp ./local/rc.sysinit          /etc/rc.d

cp ./busybox-1.00/busybox /bin
ln -s /bin/busybox /bin/sh
ln -s /bin/busybox /bin/cat
ln -s /bin/busybox /bin/ls
ln -s /bin/busybox /bin/more
ln -s /bin/busybox /bin/mount

ln -s /bin/busybox /sbin/ifconfig
ln -s /bin/busybox /sbin/init
ln -s /bin/busybox /sbin/reboot
ln -s /bin/busybox /sbin/route
ln -s /bin/busybox /sbin/ping

chmod 755 /bin /sbin
# End of File
```

Listing 13-1

Bluecat, at least the way we've built it, doesn't make use of any of these commands. So we won't try building TinyLogin into our Bluecat target. Instead, you can play around with its functionality on your workstation.

Untar tinylogin-1.4.tar.bz2 into your home directory. TinyLogin is indeed "tiny," consisting of approximately 70 source files plus documentation. The Makefile doesn't have the usual config targets. Instead, the relatively few options are managed in config.h. Comment out options you don't want and uncomment those you do. After configuration, do make. Then, as root user do **make install**. As with BusyBox, you'll have a new directory, install/, with the appropriate binary and symbolic links.

cd to _install/bin and try out some of the commands. For example, execute **tinylogin adduser sylvia**. Enter a password for sylvia when prompted. Note that you could also simply execute **adduser sylvia**.

uClinux

Now suppose you need to build an embedded product on a processor that doesn't support protected memory. And that's because your consumer electronics product can't reach its price point with a protected memory processor. There are in fact a number of microcontrollers that have the memory space capacity to run Linux but lack the memory protection mechanisms. These are available at price points that make them attractive for a number of consumer electronic devices like cell phones, digital cameras, and portable media players, to name a few.

uClinux (pronounced *you see Linux*) is a port of the Linux kernel that works without memory protection and virtual memory. It started out life as a separate source tree based on both the 2.2 and 2.4 series kernels. It was subsequently folded into the mainline source tree and is now a configuration option in the 2.6 series.

uClinux has found its way into a wide range of products including network appliances such as wireless routers, set-top boxes, industrial controllers, and even camcorders. Ports are available for the following architectures:

- Motorola m68k
- ARM7
- ETRAX
- I960
- Hitachi H8
- SPARC
- OpenRISC 1000
- MIPS

To really play around with uClinux, you'll need a target board based on one of the supported architectures. Several of these are listed in the resources section at the end of the chapter. One that I've worked with is the uCdimm from Arcturus Networks (Figure 13-3). We'll go through the process of setting up the uCdimm with uClinux as it's fairly typical of what's required for many target boards.

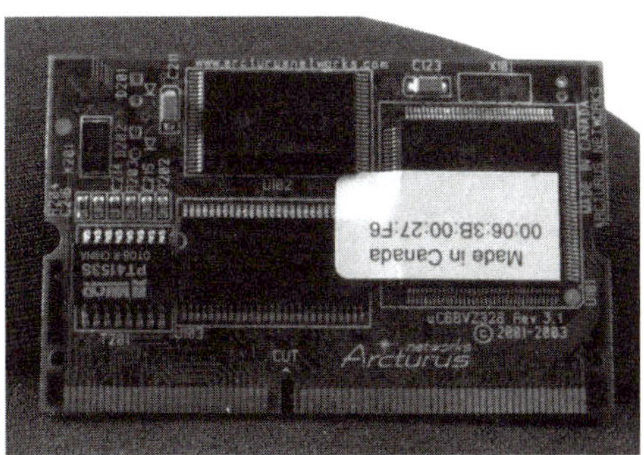

Figure 13-3: uCdimm Module

The uCdimm is based on the Motorola 68K Dragonball processor and comes with 8 Mbytes of RAM, 2 Mbytes of flash, Ethernet, an LCD controller, serial port, and miscellaneous digital I/O. To be useful, the uCdimm needs to plug into some sort of baseboard that makes its I/O readily accessible. Arcturus supplies a baseboard called the *uCEvolution*.

Installing uClinux

Although the uClinux patches are now folded into the mainline kernel source tree, that's not necessarily the best way to get started playing with it. That's because even after you select an alternate architecture like the 68k, you'll still be faced with the daunting number of configuration options, most of which you don't want in your embedded kernel image. It's easier to start with a specific uClinux distribution based on a 2.4 kernel.

The book CD has a directory `uClinux/` with three archive files in it. As root user, untar `m68k-elf-tools-incl-gdb.tar.gz` to the root, "/", directory. This installs a set of M68k cross development tools, including GDB, into directory `/opt`. `uClinux-2.4.x.tar.gz` is a 2.4.22 kernel patched for uClinux. `uClinux-dist.tar.gz` is a *very* large collection of some 35,000 files including sources for "userland" shell utilities, glibc, uClibc, documentation, and vendor information for individual target boards.

Add `/opt/uClinux-m68k-elf/bin` to the front of your $PATH. In `uClinux-dist/` execute:

```
ln -s ../uclinux-2.4.x linux-2.4.x
```

Configuring the Hardware

The uCdimm setup is almost identical to what we did with BlueCat earlier. The workstation and the target board communicate through Ethernet and a serial port. The only difference is the target's default serial baud rate is 19.2 kbps. As root user, run **minicom bluecat** (or whatever name you used for the minicom configuration), change the baud rate to 19,200 and save the configuration as, for example, **ucdimm**.

The network configuration stays the same. The workstation is 192.168.1.11 and the target is 192.168.1.200. We'll need to add another entry in the **exports** file in /etc. Add /home/<your_user_name>/uClinux-dist to **exports**. To have the new export recognized, execute:

```
/etc/rc.d/init.d/nfs stop
/etc/rc.d/init.d/nfs start
```

to cause NFS to reread the exports file.

Connect your uCdimm module to the workstation serial port and network, and power it up. You'll see something like Figure 13-4 in the minicom shell window. This is the uCdimm's bootloader. It provides facilities for loading images to flash or RAM, executing the flash or RAM images, reading and writing memory, and reading and writing bootloader environment variables. The full set of bootloader commands is found in the reference guide that came with your uCdimm module.

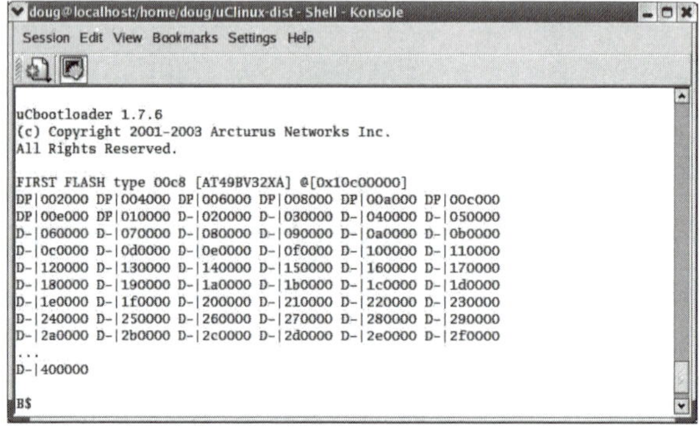

Figure 13-4: uCbootloader Initial Screen

Try the command **printenv**. This results in the output shown in Figure 13-5[2]. The bootloader's environment variables provide a way to customize the module without having to rebuild the kernel. In particular, we can specify an NFS mount point. Execute:

```
setenv NFSMOUNT=192.168.1.11:/home/<your_user_name>/uClinux-dist
```

to create a mount point that matches what you exported above.

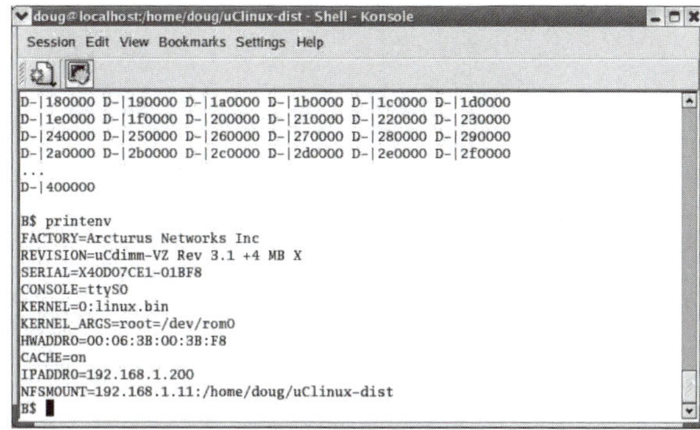

Figure 13-5: Bootloader Environment Variables

Unfortunately, to go any farther with this, we'll have to execute a uClinux build. For now though, execute the bootloader's **go** command to boot up the default Linux image. When the kernel boot completes, you'll get a login prompt. The default user is "root," the password is "uClinux." A successful login returns the shell's "#" prompt. Now you can play around with the commands built into uClinux, many of which come from BusyBox.

Configuring and Building uClinux

We need to go through the build process to get a ROM filesystem that we can modify. The build procedure insists that you do a configuration first, even though all of the defaults are perfectly OK. In **uClinux-dist/** execute **make menuconfig** (uClinux doesn't support **xconfig**). Your only real choice here is "Target Platform Selection." Select that to bring up the menu in Figure 13-6. Select "Vendor/Product." This lets you specify a vendor and possibly one of several boards from that vendor so that appropriate configuration options are automatically selected. Note that the default is correct. The only way to exit the Vendor/Product list is with Select.

---

[2] Well, almost. The NFSMOUNT variable will have a different value.

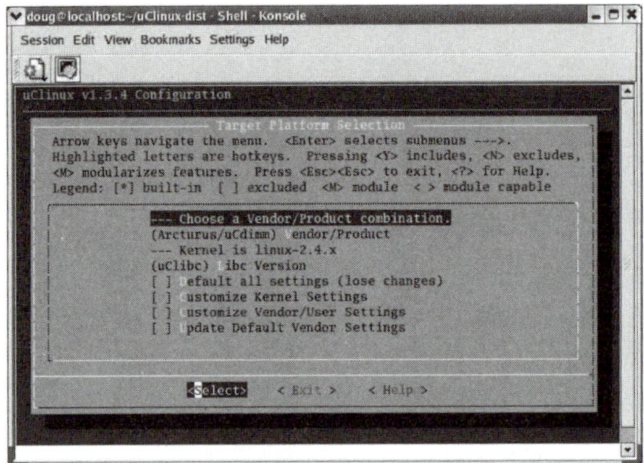

Figure 13-6: uClinux configuration

The "Libc Version" option lets you select which version of the standard C library will be included in your target filesystem. Again, the default uClibc is a good choice. Enabling the "Customize Kernel Settings" option gives you the chance to configure the kernel. Upon exiting the top-level uClinux menu, the kernel's **menuconfig** will be invoked. For now the defaults will suffice.

Exit all the way out and say yes when asked if you want to save the new "kernel" configuration. The first time you exit from **menuconfig** it builds the **.config** file. Next run **make dep**. This is a step required by 2.4 and earlier kernels to build a set of *dependency files* that show what depends on what. Now execute **make** to build a flash loadable image containing the target kernel and root filesystem.

Following the build you'll find two new directories under **uClinux-dist/**. **images/** contains, among other things, **image.bin** that gets written to the target's flash. It's a concatenation of the kernel, **linux.txt**, with the root filesystem, **romfs.img**. **romfs/** is the contents of the root filesystem. Its subdirectories mimic the kernel's standard file hierarchy.

Open **romfs/etc/rc** with an editor. Down around line 13 you'll see several lines such as:

```
printbenv –q –e IPADDR0 >> /etc/profile
```

This is a way of importing the bootloader environment variables into the kernel. Near the end of the **rc** file is the line:

```
mount $NFSMOUNT
```

Hmmm...Given the way we defined NFSMOUNT previously, something seems to be missing, specifically "–t nfs" and a mount point in the local filesystem. That's because there's a bug in the bootloader such that it will not accept spaces in environment variables. The solution is to change the above line to:

```
mount –t nfs $NFSMOUNT /usr
```

Make the change and save the file.

Downloading a Target Image

Back in `uClinux-dist/` execute `make image`. This creates a new target image with the revised `/etc/rc` file. Now we need to load that image file into the target's flash memory. Since we don't yet have NFS access to the workstation's filesystem, we'll have to use the serial port.

Press the Reset button on the target to get back to the bootloader. We could load the image at the default serial baud rate of 19,200, but that would take a long time. The bootloader offers a speedy alternative for downloading large image files. The `fast` command changes the serial port bit rate to 115,200 bps. In the minicom shell window do the following:

- Enter fast to the bootloader;
- Type Ctrl + A O to get to configure minicom and select Serial port setup;
- Select "E", then "I" to change to 115,200 bps;
- Exit back out without saving and type <Enter> a couple of times to verify that you have the bootloader prompt.

Now we're ready to download our new target image. Execute the bootloader's `rx` command. Type Ctrl + A S (Send file) and select the Xmodem protocol. Move to the images directory and select `romfs.img` as the file to download. This copies the file into the target's RAM. When the download is complete, execute the bootloader's program command to move the image into flash.

When the program operation finishes, execute `slow` to return the serial port to 19.2 kbps and do the corresponding operation in minicom. Now execute `go` and the new image will boot and mount your NFS volume.

You're now ready to experiment with uClinux. Create a subdirectory under uClinux-dist/ for your own code. Try the thermostat example from Chapter 6. Since the target is not a protected mode environment, there's nothing to prevent a normal process from executing I/O. A device driver can be just a set of functions linked to the application, either statically or dynamically. Implement a driver for the pushbuttons and LEDs on the uCdimm to support the thermostat.

uClinux includes gdbserver so you can remotely debug your code using GDB and DDD.

Target Boards in General

The uCdimm is fairly typical of most SBCs (single board computer) that are capable of running Linux. Whatever architecture and board you choose for your project, the process of getting it up and running with your application generally follows the steps outlined above.

Most SBCs have at least 8 Mbytes of RAM and a few megabytes of flash, sometimes in the form of a CompactFlash module, enough to hold an image containing a Linux kernel and a root filesystem. A bootloader supports downloading and programming an image through a serial port or over the network, and subsequently booting that image.

The board will support a variety of I/O including Ethernet, serial, USB, a parallel port and a range of digital I/O. In many cases an IDE port is available for attaching disk drives.

Building uClinux for the 2.6 Kernel

This section uses uClinux as an example of how to build a kernel for an architecture other than that of the workstation. This is known as *cross compiling*. Remember from chapter 4 that the configuration menu resides in linux/arch/($ARCH)/ where ARCH is a Makefile variable. ARCH normally defaults to the architecture of the host workstation. To build the kernel for a different architecture, we simply have to change the value of ARCH.

Again, before experimenting around with new kernel configurations, it's a good idea to copy the source tree into a new directory under /usr/src. In that new directory execute:

```
make ARCH=m68knommu xconfig
```

make now gets the configuration menu from the m68knommu/ directory under arch/. You'll see a totally different set of options in the "Processor type and features" submenu. The default processor is the M68VZ328. The platform is mislabeled uD-simm but really does refer to the uCdimm.

Before actually building the M68K kernel, we need to make a simple change in the kernel source tree. In the linux/include/ directory there's a link called asm that currently points to asm-i386/. Delete that link and make a new one that points to asm-m68knommu/. You'll probably also want to change the EXTRAVERSION string in the Makefile to reflect the new kernel, perhaps "uclinux." Now build the M68K kernel with:

```
make ARCH=m68knommu CROSS_COMPILE=m68k-elf-
```

Note the "-" at the end. The CROSS_COMPILE string is prepended to all the build tools such as the compiler, assembler, linker, and so on. Remember that our M68K compiler is called m68k-elf-gcc. Of course, the cross tools must be in your PATH.

Rather than entering the ARCH and CROSS_COMPILE arguments at runtime, you could export them as environment variables. This is the preferred solution once you're working with a specific target. You could also edit the Makefile to change them permanently, but that's probably not a good idea.

Of course the kernel is only half the problem. The other half is building a compressed root filesystem to load into flash. The kernel build system doesn't provide any resources for that. Perhaps the best way to learn about building the filesystem is to look at the Makefile in uClinux-dist/vendors/Arcturus/uCdimm/. The romfs: target shows how the filesystem is constructed. The image: target shows how the resulting filesystem is compressed into a single file using genromfs and objcopy. See the man pages for these utilities to learn how they work. Finally, the filesystem is concatenated with the kernel to yield the image that is loaded into flash.

Summary

This chapter has been a quick tour of some of the many tools and resources that are out there for building embedded systems with Linux. Each of the topics we've looked at: BusyBox, TinyLogin, and uClinux deserve additional study and investigation.

Having come to the end of our ride through the "twisty little passages known as *Linux*." you should have a fairly good grasp of how Linux is structured, how the tools work in general, and how and where to find additional information. You're ready to dive in and figure this stuff out on your own. Good luck and have fun!

But before we sign-off entirely, there's one more topic we should take a look at because it's having an increasingly significant impact on embedded software development in both the Open Source and commercial worlds. The next, and last chapter, explores the Eclipse integrated development environment.

Resources

www.busybox.net – This is the official BusyBox website. Download the latest version, join a mailing list, or visit the "Hall of Shame" identifying vendors who use BusyBox but apparently don't release source code as required by the GPL.

http://tinylogin.busybox.net/ – The official website for Tinylogin. Very similar to the BusyBox website. Not surprising since both packages are maintained by the same person, Erik Andersen.

www.uclinux.org – The principal website for uClinux.

There are many sources of target boards for experimenting with uClinux. Here is a random sampling:

www.arcturus.com – Arcturus Networks makes the uCdimm module.

www.steroidmicros.com – Steroid Micros offers a development kit based on the 5282 Coldfire processer.

www.simtec.co.uk – Simtec offers a tiny board-level computer based on a 60 MHz ARM7 processor with uClinux preinstalled.

www.ampro.com – Ampro practically invented the single board computer business. They developed the PC/104 standard form factor for SBCs that is now widely used in embedded systems. Ampro makes boards based on the x86 in several form factors in addition to PC/104.

www.arcom.com – Arcom supplies a wide range of PC/104 SBCs based on the Intel x86 and various XScale[3] processors. They also offer complete development kits with Linux preinstalled. These are a bit pricey, but if you can afford it, they're a quick way to get up and running.

[3] An implementation of the ARM architecture.

CHAPTER 14

Eclipse Integrated Development Environment[1]

Integrated development environments (IDE) are a great tool for improving productivity. Desktop developers have been using them for years. Perhaps the most common example is Microsoft's Visual Studio environment. In the past, a number of embedded tool vendors have built their own proprietary IDEs.

In the Open Source world, the IDE of choice is Eclipse, also known as the *Eclipse Platform* and sometimes just the *Platform*. The project was started by Object Technology International (OTI), which was subsequently purchased by IBM. In late 2001, IBM and several partners formed an association to promote and develop Eclipse as an Open Source project. It is now maintained by the Eclipse Foundation, a non-profit consortium of software industry vendors. Several leading embedded Linux vendors such as Monta Vista, TimeSys, LinuxWorks, and Wind River Systems have embraced Eclipse as the basis for their future IDE development.

It should be noted that Eclipse is not confined to running under Linux. It runs just as well under Windows 2000 and XP.

Overview

"Eclipse is a kind of universal tool platform—an open, extensible IDE for anything and nothing in particular. It provides a feature-rich development environment that allows the developer to efficiently create tools that integrate seamlessly into the

---

[1] Portions of this chapter are adapted from Eclipse Platform Technical Overview, © IBM Corporation and The Eclipse Foundation, 2001, 2003, 2005, and made available under the Eclipse Public License (EPL). The full document is available at *www.eclipse.org*.

Eclipse platform," according to the platform's own on-line overview. Technically, Eclipse itself is not an IDE, but is rather an *open platform* for developing IDEs and *rich client* applications.

Figure 14-1 show the basic Eclipse workbench. It consists of several *views* including:

- *Navigator* – shows the files in the user's workspace
- *Text Editor* – shows the contents of a file
- *Tasks* – a list of "to dos"
- *Outline* – of the file being edited. The contents of the outline view are content-specific

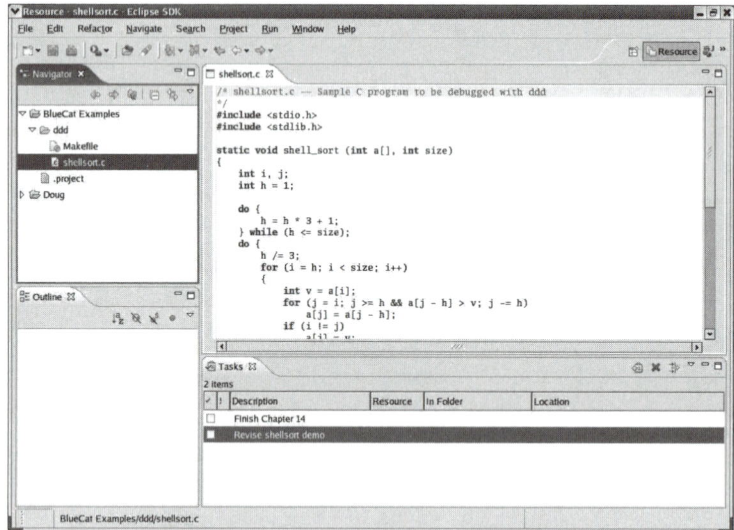

Figure 14-1: Eclipse Workbench

Although Eclipse has a lot of built-in functionality, most of that functionality is very generic. It takes additional tools to extend the platform to work with new content types, to do new things with existing content types, and to focus the generic functionality on something specific.

The Eclipse Platform is built on a mechanism for discovering, integrating, and running modules called *plug-ins*. A tool provider writes a tool as a separate plug-in that operates on files in the workspace and exposes its tool-specific UI in the workbench. When you launch Eclipse, it presents an integrated development environment (IDE) composed of the set of available plug-ins.

The Eclipse platform is written primarily in Java and in fact was originally designed as a Java development tool. Figure 14-2 shows the platform's major components and APIs. The platform's principal role is to provide tool developers with mechanisms to use, and rules to follow, for creating seamlessly integrated tools. These mechanisms are exposed via well-defined API interfaces, classes, and methods. The platform also provides useful building blocks and frameworks that facilitate developing new tools.

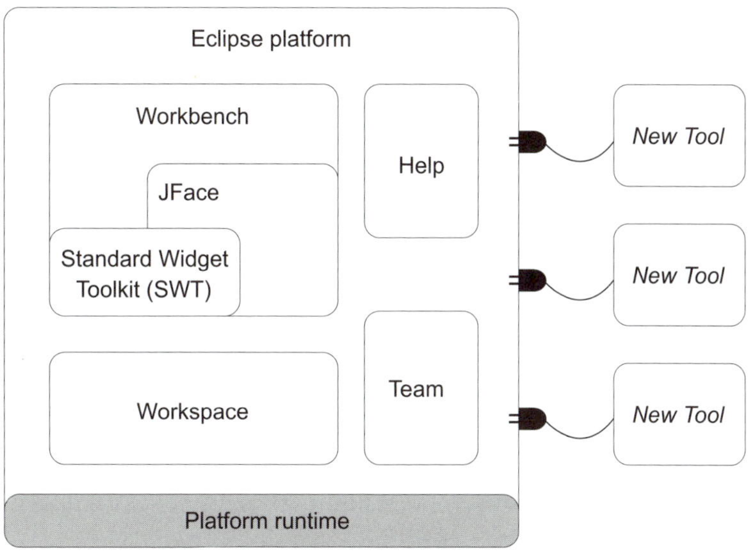

Figure 14-2: Eclipse Platform Architecture

Eclipse is designed and built to meet the following requirements:

- Support the construction of a variety of tools for application development.
- Support an unrestricted set of tool providers, including independent software vendors (ISVs).
- Support tools to manipulate arbitrary content types such as HTML, Java, C, JSP, EJB, XML, and GIF.
- Facilitate seamless integration of tools within and across different content types and tool providers.
- Support both GUI and non-GUI-based application development environments.
- Run on a wide range of operating systems, including Windows and Linux.

- Capitalize on the popularity of the Java programming language for writing tools.

Plug-ins

A *plug-in* is the smallest unit of Eclipse functionality that can be developed and delivered separately. A small tool is usually written as a single plug-in, whereas a complex tool may have its functionality split across several plug-ins. Except for a small kernel known as the *Platform Runtime*, all of the Eclipse platform's functionality is located in plug-ins.

Plug-ins are coded in Java. A typical plug-in consists of Java code in a JAR library, some read-only files, and other resources such as images, web templates, message catalogs, native code libraries, etc. Some plug-ins don't contain code at all. An example is a plug-in that contributes online help in the form of HTML pages. A single plug-in's code libraries and read-only content are located together in a directory in the file system, or at a base URL on a server.

Each plug-in's configuration is described by a pair of files. The *manifest* file, manifest.mf, declares essential information about the plug-in, including the name, version, and dependencies to other plug-ins. The second optional file, plugin.xml, declares the plug-in's interconnections to other plug-ins. The interconnection model is simple: a plug-in declares any number of named *extension points*, and any number of *extensions* to one or more extension points in other plug-ins.

The extension points can be extended by other plug-ins. For example, the workbench plug-in declares an extension point for user preferences. Any plug-in can contribute its own user preferences by defining extensions to this extension point.

On start-up, the Platform Runtime discovers the set of available plug-ins, reads their manifest files, and builds an in-memory plug-in registry. The platform matches extension declarations by name with their corresponding extension point declarations. Any problems, such as extensions to missing extension points, are detected and logged. The resulting plug-in registry is available via the Platform API. After startup, plug-ins can be unloaded, and new ones installed or new versions of existing plug-ins can replace existing versions.

A plug-in is *activated* when its code actually needs to be run. Once activated, a plug-in uses the plug-in registry to discover and access the extensions contributed to its extension points. For example, the plug-in declaring the user preference extension point can discover all contributed user preferences and access their display names to construct a preference dialog. This can be done using only the information from the registry, without having to activate any of the contributing plug-ins. The contributing plug-in will be activated when you select one of its preferences from a list.

By determining the set of available plug-ins up front, and by supporting a significant exchange of information between plug-ins without having to activate any of them, the platform can provide each plug-in with a rich source of pertinent information about the context in which it is operating. The context doesn't change while the platform is running, so there's no need for complex life cycle events to inform plug-ins when the context changes. This avoids a lengthy start-up sequence and a common source of bugs stemming from unpredictable plug-in activation order.

Workspaces

The various tools plugged in to the Eclipse Platform operate on regular files in your *workspace*. The workspace consists of one or more top-level *projects*, where each project maps to a corresponding directory in the file system. The different projects in a workspace may map to different file system directories or drives, although, by default, all projects map to sibling subdirectories of a single workspace directory.

Each project contains files that you create and modify. All files in the workspace are directly accessible to the standard programs and tools of the underlying operating system. Tools integrated with the Platform are provided with APIs for dealing with workspace *resources* (the collective term for projects, files, and folders). So-called *adaptable objects* represent workspace resources so that other parties can extend their behavior.

In a large project, the Linux kernel for example, tools like compilers and link checkers must apply a coordinated analysis and transformation of thousands of separate files. To this end the platform provides an *incremental project builder* framework; the input to an incremental build is a *resource tree delta* capturing the net resource differences since the last build. Sophisticated tools may use this mechanism to provide scalable solutions. The platform allows several different incremental project builders to be registered on the same project and provides ways to trigger project and workspace-wide builds. An optional workspace auto-build feature automatically triggers the necessary builds after each resource modification operation (or batch of operations).

Workbench

The Eclipse *workbench* API, user interface (UI), and implementation are built from two toolkits:

- Standard Widget Toolkit (SWT) - a widget set and graphics library integrated with the native window system but with an OS-independent API.
- JFace – a UI toolkit implemented using SWT that simplifies common UI programming tasks.

Unlike SWT and JFace, which are both general purpose UI toolkits, the workbench provides the UI personality of the Eclipse Platform, and supplies the structures in which tools interact with the user. Because of this central and defining role, the workbench is synonymous with the Eclipse Platform UI as a whole and with the main window you see when the Platform is running (see Figure 14-1). The workbench API is dependent on the SWT API, and to a lesser extent on the JFace API.

The Eclipse Platform UI paradigm is based on editors, views, and perspectives. From the user's standpoint, a workbench window consists visually of views and editors. Perspectives manifest themselves in the selection and arrangements of editors and views visible on the screen.

Editors allow the user to open, edit, and save objects. They follow an open-save-close lifecycle much like file system based tools, but are more tightly integrated into the workbench. When active, an editor can contribute actions to the workbench menus and tool bar. The platform provides a standard editor for text resources; other plug-ins supply more specific editors.

Views provide information about some object that you are working with. A view may assist an editor by providing information about the document being edited. For example, the standard content outline view shows a structured outline for the content of the active editor if one is available. A view may augment other views by providing information about the currently selected object. For example, the standard properties view presents the properties of the object selected in another view. The platform provides several standard views; additional ones are supplied by other plug-ins.

A workbench window can have several separate *perspectives*, only one of which is visible at any given moment. Each perspective has its own views and editors that are arranged (tiled, stacked, or detached) for presentation on the screen although some may be hidden. Several different types of views and editors can be open at the same time within a perspective. A perspective controls initial view visibility, layout, and action visibility. You can quickly switch perspective to work on a different task, and can easily rearrange and customize a perspective to better suit a particular task. The platform provides standard perspectives for general resource navigation, online help, and team support tasks. Other plug-ins supply additional perspectives.

Plug-in tools may augment existing editors, views, and perspectives to:

- Add new actions to an existing view's local menu and tool bar.
- Add new actions to the workbench menu and tool bar when an existing editor becomes active.
- Add new actions to the pop-up content menu of an existing view or editor.
- Add new views, action sets, and shortcuts to an existing perspective.

The platform takes care of all aspects of workbench window and perspective management. Editors and views are automatically instantiated as needed, and disposed of when no longer needed. The display labels and icons for actions contributed by a tool are listed in the plug-in manifest so that the workbench can create menus and tool bars without activating the contributing plug-ins. The workbench doesn't activate the plug-in until you try to use functionality that the plug-in provides.

Installation

There are two parts to Eclipse, both in the `/tools` directory on the CD. `eclipse-SDK-3.1.2-linux-gtk.tar.gz` is the platform itself from Eclipse.org. You can untar this just about anywhere you like. I chose to put it in my home directory, although that may not be the optimal choice. After untarring, you'll find a new subdirectory called, not surprisingly, `eclipse/`.

Since Eclipse is written primarily in Java, you'll need a Java runtime environment (JRE) to support it. `jre-1_5_0_06-linux-i586.bin` is the JRE from Sun Microsystems. This is a binary executable. Copy it to `/usr/local` and execute it. This creates a new subdirectory `jre1.5.0_06/`. You'll need to add `/usr/local/jre1.5.0_06/bin` to your PATH.

That's it! Eclipse is installed and ready to go. In a shell windows, `cd` to the `eclipse/` directory and execute `./eclipse`. Or just double-click it in a KDE file manager window.

Using Eclipse

When Eclipse first starts up, it asks you to select a workspace. The default is called "workspace" in the directory where you installed Eclipse. Following the workspace dialog, if this is the first time you've executed Eclipse, you'll see the Welcome screen shown in Figure 14-3. On subsequent runs Eclipse opens directly in the workbench, but you can always access this window from the first item on the Help menu, Welcome.

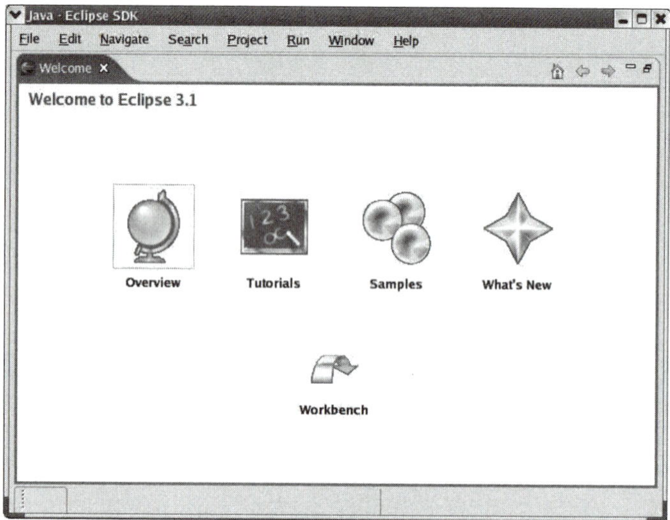

Figure 14-3: Eclipse Welcome Screen

The Tutorial icon leads you to a very good, very thorough introduction to using Eclipse. If you're new to the Eclipse environment, I strongly suggest you go through it. There's no point in duplicating it here. Go ahead, I'll wait.

OK, now you have a pretty good understanding of generic Eclipse operation. It's time to get a little more specific and apply what we've learned to C software development for embedded systems.

The C Development Environment (CDT)

Among the ongoing projects at Eclipse.org is the tools project. A subproject of that is the C/C++ Development Tools (CDT) project whose aim is to build a fully functional C and C++ integrated development environment (IDE) for the Eclipse platform. CDT under Linux is based on GNU tools and includes:

- C/C++ Editor with, syntax coloring
- C/C++ Debugger using GDB
- C/C++ Launcher for external applications
- Parser
- Search Engine
- Content Assist Provider
- Makefile generator

Installing CDT

Installing CDT is almost as easy as installing Eclipse itself. You need to be connected to the Internet to use this process. Click on **Help -> Software Updates -> Find and Install**. In the **Feature Updates** window select **Search for new features to install** and click **Next**. Click the **New Remote Site** button and enter the location of the CDT install site:

```
http://download.eclipse.org/tools/cdt/releases/eclipse3.1
```

If that's too much to type, go to *www.eclipse.org/tools*, find the link under Current Releases and copy it. Now select the update site you just created by clicking the appropriate checkbox and click **Next**.

A dialog box appears showing the updates available from this update site, select each of the following features, ensuring you have precisely the same version for each one:

- Eclipse C/C+ Development Tooling SDK
- Eclipse C/C+ Development Tools

Then click **Next**. Now you see the Eclipse.org Software User Agreement, which you must accept to install the CDT update. Select the location you where you want the updates installed, by default the directory where Eclipse is installed, and click **Finish**. After the download completes, click **Install** for each of the downloaded features. Finally, you'll need to restart Eclipse for the CDT to be recognized.

Creating a Project

Typically, when you start working with CDT, you've already got a number of projects put together in the "conventional" way such as the exercises in this book. Or you might have projects managed by a CVS repository. So you'd probably like to bring one or more of those projects into the Eclipse CDT environment.

Let's use the thermostat project as an example. In Eclipse, select **File → New → Project**. That brings up the window shown in Figure 14-4. Select **Standard Make C Project** and click **Next**. In the next window (Figure 14-5), enter the project name, "thermostat," and uncheck the **Use default** box. Now browse to the **thermostat/** directory under **Bluecat/**. Click Finish. The project wizard creates the project and then asks if you'd like to open the C/C++ perspective. Yes, you would.

Down at the bottom of the screen is a tab labeled Problems that says "*** No rule to make target 'all'. That's true. The thermostat Makefile does not in fact have an "all" target. Turns out that by default, the only thing CDT knows how to make is "all".

Open the Makefile. Note first of all that the Outline tab shows all of the macros and targets defined in the Makefile. Now just above the line that says "sim : thermostat.s devices" add a line "all : sim". For our purposes we'll just build the simulation version of thermostat. Save the Makefile and select **Project → clean...** and check thermostat and click OK. The project will be rebuilt.

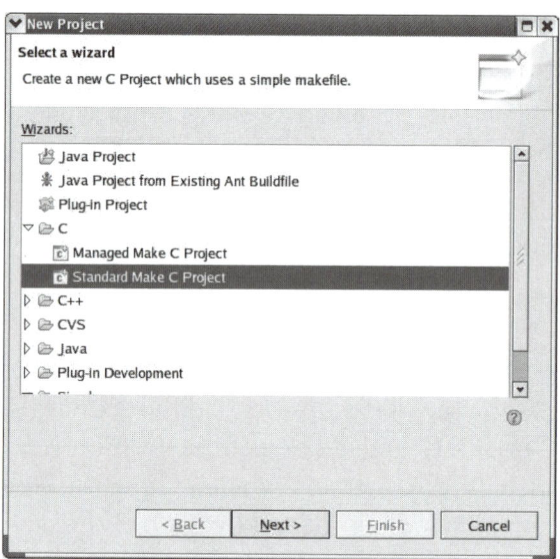

Figure 14-4: New Project Wizard

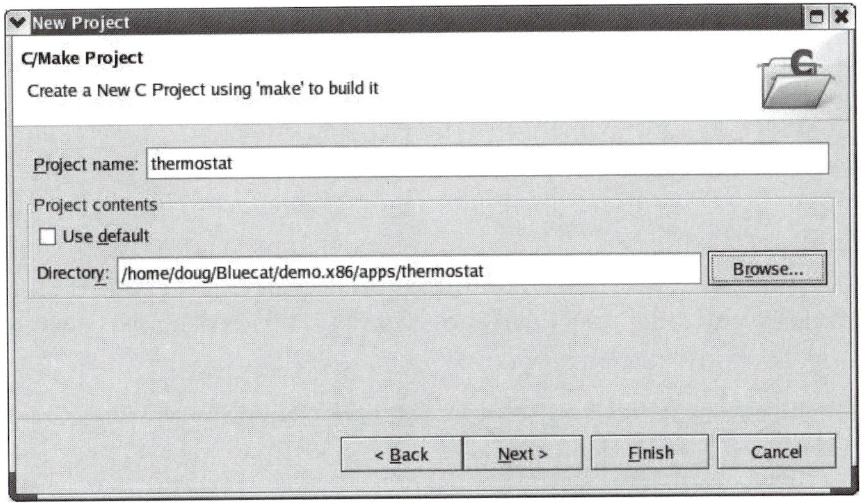

Figure 14-5: New Project Wizard, Part 2

Debugging with CDT

CDT very nicely integrates GDB in a manner similar to DDD. Remember that the simulated thermostat relies on another program called devices, which was also made by the build process above. Before debugging thermostat.s, we need to start devices in a shell window. Go ahead and do that.

The first time you debug a project, you'll need to set up a debug launch configuration as follows:

1. In C/C++ Projects view, select the thermostat project.
2. Click Run → Debug.
3. In the Debug dialog box, select C/C++ Local Application from the Configurations list.
4. Click New. The new configuration is named thermostat by default and refers to the thermostat project
5. In the C/C++ Application box, type the name of the executable that you want to run, thermostat.s.
6. Click the Debugger tab.
7. In the Debugger drop-down menu, select GDB debugger.

8. Make sure **Stop at main() on startup** is checked.
9. Click **Debug**.

Eclipse will ask if you want to change to the Debug perspective. Yes, you do. Figure 14-6 shows the initial Debug perspective. Down at the bottom is a Console window where terminal I/O takes place. The Editor window shows **thermostat.c** with the first line of **main()** highlighted. That's because we said Stop at **main()** in our debug configuration setup. The Outline tab on the right lists all of the header files included by **thermostat.c** and all of the global and external symbols declared in the file including variables, constants, and functions.

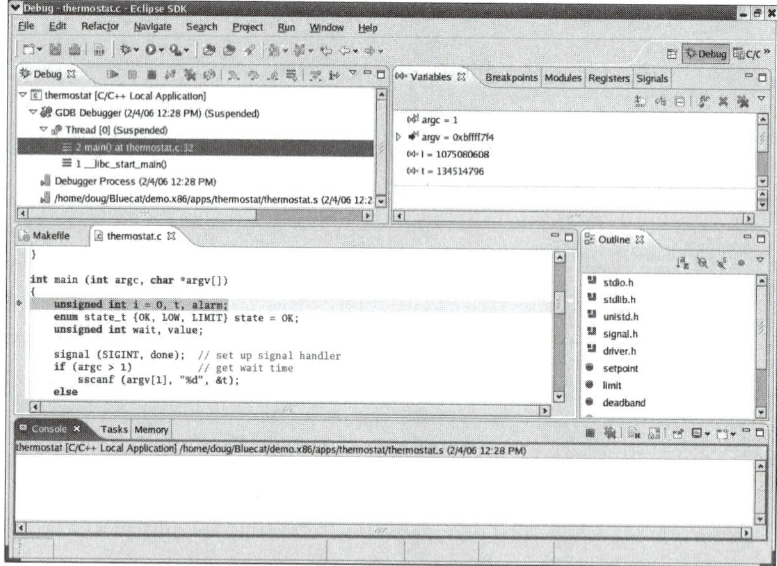

Figure 14-6: Debug Perspective

In the upper right is a set of tabs that provide a range of debug information including local variables, breakpoints, and registers. Initially of course all the local variables have random values since they're allocated on the stack. Just above the Debug tab in the upper left is a set of execution icons. Click on the Step Over icon. The highlight in the Editor window advances to the next line and i in the Variable tab show its initial value of 0. It's shown in red because it changed since the last time the debugger was suspended.

In the Editor window, move down to the **while(running)** line and right click. Select **Run to line**. The program executes up to this line and stops. Note that most of the local variables have now been initialized. As with DDD, we can simply move the cursor over a variable name to see its current value.

Working with Breakpoints

Go to the line that says `i++`. Right click in the frame on the left side of the Editor and select `Toggle Breakpoint`. The green circle indicates that an enabled breakpoint is set on this line. The checkmark simply indicates that the breakpoint was successfully installed. Open the Breakpoints tab to see that the breakpoint is listed there. The checkbox indicates whether or not the breakpoint is enabled. Click it to disable the breakpoint and notice that the green circle becomes clear to indicate a disabled breakpoint. Click the checkbox again to re-enable the breakpoint.

Click the `Resume` button and the program executes to the breakpoint. The Console window now shows the first output from the program. So far, so good, but chances are we don't really want to break every time through the loop. We only need to break when some condition changes in a way that alters the program's behavior.

Go to the Breakpoint tab, right click on our breakpoint and select Properties. This gives us the option to qualify the breakpoint on a condition. Let's set the condition to be `value < 32` so the program stops when we expect the heater to come on. Note that the other thing we can do is set an "ignore count". Setting this to a non-zero value causes the breakpoint to be ignored that many times before stopping the program.

Close the Properties dialog and note that the condition has been added to the breakpoint description. Click Resume and watch the program print a few values. Now go to the devices program, which should still be running, and set the A/D value to 29 or less. When the program hits the breakpoint, single step to verify that state changes to `LOW` and the heater turns on. At this point you probably want to change the breakpoint condition to be `value > 32` so you can watch what happens when the heater is supposed to turn off. You could also set the condition to be `value > limit` to verify that path of the program.

When you're finished click the `Terminate` button to stop the debugger. Change back to the C/C++ perspective.

Remote Debugging

Now let's move on to the embedded target machine. Eclipse CDT supports GDB server so we can debug remotely just as we did earlier with DDD. First we need to change the CDT build target to build the "target" version of thermostat. In the C/C++ Projects tab select the thermostat project and right click. Way down at the bottom select Properties. Then select C/C++ Make Project to bring up the dialog box shown in Figure 14-7.

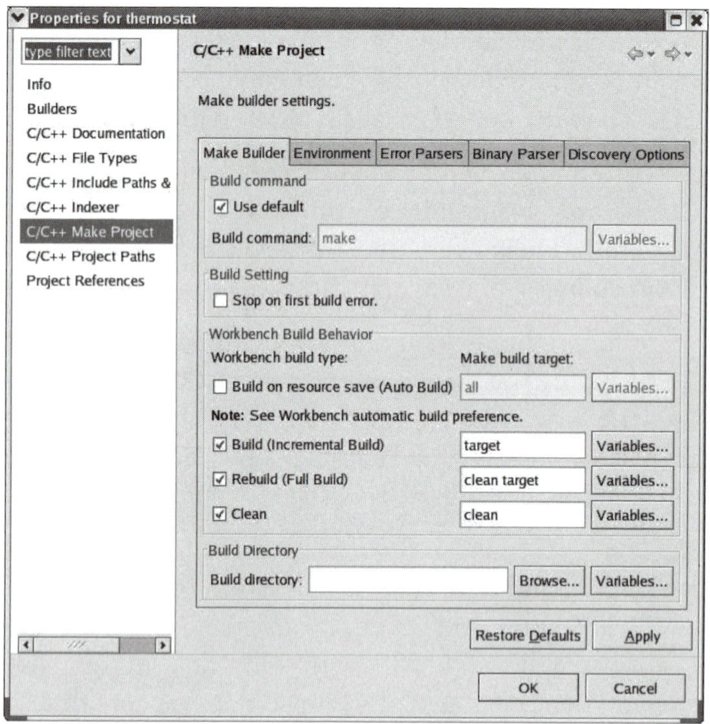

Figure 14-7: Project Properties, Make Project

For the Build and Rebuild entries change "all" to "target". When you click OK the project will be rebuilt, this time using the "target" target.

Fire up your embedded target and boot the shell kernel. Execute the following commands to the target:

```
cd /usr/demo.x86/apps/thermostat
gdbserver :1234 thermostat.t
```

Back in Eclipse, change to the Debug perspective and select **Run -> Debug...** Click **New** to create a new debug configuration. Name it "remote". The project is still "thermostat" but the application is now "**thermostat.t**". In the Debugger tab (Figure 14-8) select GDB Server. Under Connection select TCP and set the Host IP address to 192.168.1.200 and the port to 1234. Click **Debug**. You're now debugging on the target and able to do everything we did earlier debugging on the workstation.

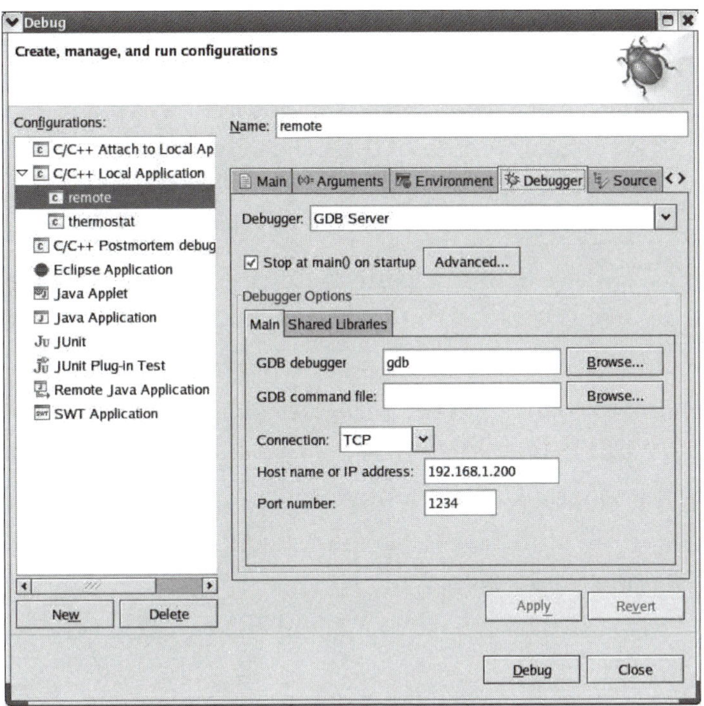

Figure 14-8: Debug Configuration, GDB Server

Summary

In my humble opinion, Eclipse is the most professionally executed Open Source project I've seen. It's well thought out and meticulously implemented. While it has tremendous power and flexibility, the most common operations are relatively simple and intuitive. The documentation, while still reference in nature and not tutorial, is nevertheless readable, and for the most part accurate. Sadly, there are all too many Open Source projects for which that can't be said.

The plug-in mechanism makes Eclipse infinitely extensible. This opens up opportunities for both Open Source and commercial extensions to address a wide range of applications. The Resources section below lists a couple of repositories of Eclipse plug-ins.

Like many of the topics in this book, we've really just skimmed the surface, focusing on those areas that are of specific interest to embedded developers. But hopefully this has been enough to pique your interest in digging further. There's a lot more out there. Go for it!

Resources

www.eclipse.org – The official website of the Eclipse Foundation. There's a lot here and it's worth taking the time to look through it. I particularly recommend downloading the C/C++ *Development Toolkit User Guide* (PDF).

http://eclipse-plugins.2y.net – Eclipse Plugins. This site will give you a feel for the extent of the Eclipse ecosystem. It lists over 1000 plug-ins, both commercial and Open Source.

www.eclipseplugincentral.com – Eclipse Plugin Resource Center and Marketplace. Not quite as extensive as Eclipse Plugins, this site lists some 400 plug-ins.

A Google search on "eclipse plug-in" returns a great many hits, but except for the two sites listed above, all of the others seem to describe specific plug-ins mostly oriented toward Java and web development.

APPENDIX A

RTAI Application Programming Interface (API)

This is a summary of the RTAI API derived primarily from the HTML documentation supplied with the package and also available online from rtai.org. It is organized by module.

Module: hal

#include "hal.h"

This module is the interface to Adeos. It provides services for managing interrupts between RTAI and Linux.

int rt_startup_irq (unsigned *irq*); Start and initialize the PIC to accept interrupt request *irq*. Return value not specified.

void rt_shutdown_irq (unsigned *irq*); Shut down an IRQ source. No further interrupts will be accepted for this IRQ.

void rt_enable_irq (unsigned *irq*); Enable an IRQ source.

void rt_disable_irq (unsigned *irq*); Disable an IRQ source.

Caution: These functions manipulate the PIC (programmable interrupt controller). You shouldn't use them unless you know what you're doing.

void rt_mask_and_ack_irq (unsigned *irq*); Mask and acknowledge an IRQ source. RTAI calls this function for a level-triggered interrupt before passing control to your interrupt handler.

void rt_ack_irq (unsigned *irq*); Acknowledge an IRQ source. RTAI calls this function for an edge-triggered interrupt before passing control to your interrupt handler.

void rt_unmask _irq (unsigned *irq*); Unmask an IRQ source. Call this when your level-triggered interrupt handler is finished. Not needed for edge-triggered interrupts.

int rt_request_linux_irq (unsigned int irq, void (*handler)(int irq, void *dev_id, struct pt_regs *regs), char *name, void *dev_id);

int rt_free_linux_irq (unsigned int irq, void *dev_id);

rt_request_linux_irq installs function *handler* as an interrupt service routine for IRQ level *irq* forcing Linux to share the IRQ with other interrupt handlers. The handler is appended to any already existing Linux handler for the same IRQ and run as a Linux IRQ handler. In this way a real time application can monitor Linux interrupt handling at is will. The handler appears in /proc/interrupts. *name* is a name for /proc/interrupts. The parameter *dev_id* is passed to the interrupt handler, in the same way as the standard Linux IRQ request call. The interrupt service routine can be uninstalled with rt_free_linux_irq.

Return value

- 0 – Success
- -EINVAL – *irq* is not a valid IRQ number or *handler* is NULL
- -EBUSY – There is already a handler for interrupt *irq*

void rt_pend_linux_irq (unsigned int *irq*);

Appends a Linux interrupt *irq* for processing in Linux IRQ mode, i.e., with interrupts fully enabled..

int rt_request_srq (unsigned int *label*, void (*))(void) *k_handler*, long long (*)(unsigned) *u_handler*);

int rt_free_srq (unsigned int *srq*);

int rt_request_srq installs a two way RTAI system request (srq) by assigning *u_handler*, a function to be used when a user calls srq from user space, and *k_handler*, the function to be called in kernel space following its activation by a call to rt_pend_linux_srq. *k_handler* is in practice used to request a service from the kernel. In fact Linux system requests cannot be used safely from RTAI so you can setup a handler that receives real time requests and safely executes them when Linux is running.

rt_free_srq uninstalls the system call identified by *srq*.

Return value

- 0 – Success
- -EINVAL – *k_handler* is NULL or *srq* is invalid
- -EBUSY – No free srq slot is available

void rt_pend_linux_srq (unsigned int *srq*);

Appends a system call request *srq* to be used as a service request to the Linux kernel. *srq* is the value returned by rt_request_srq().

Timer Management

void rt_request_timer (void (**handler*)(void), int *tick*, int *apic*);

void rt_free_timer (void);

rt_request_timer requests a timer of period *tick* ticks, and installs the routine *handler* as a real time interrupt service routine for the timer. Set *tick* to 0 for one-shot mode. If *apic* has a nonzero value the local APIC timer is used. Otherwise timing is based on the 8254.

rt_free_timer uninstalls the timer previously set by rt_request.

Diagnostics

int rt_printk (const char **format*, ...);

RTAI-safe version of printk().

Return value

- 0 – Number of characters printed

RTAI Scheduler Modules

The modules in this section implement the basic task and communication mechanisms of RTAI.

Task Functions

#include "rtai_sched.h"

int rt_task_init (RT_TASK **task*, void (**rt_thread*)(int), int *data*, int *stack_size*, int *priority*, int *uses_fpu*, void(**signal*)(void));

int rt_task_init_cpuid (RT_TASK **task*, void (**rt_thread*)(int), int *data*, int *stack_size*, int priority, int *uses_fpu*, void(**signal*)(void), unsigned int *cpuid*);

Create a real time task. *task* is a pointer to an RT_TASK type structure whose space must be provided by the application. It must be kept during the whole lifetime of the real time task and cannot be an automatic variable.

rt_thread is the entry point of the task function. The parent task can pass a single integer value *data* to the new task.

stack_size is the size of the stack to be used by the new task, and *priority* is the priority to be given the task. The highest priority is 0, while the lowest is RT_LOWEST_PRIORITY.

uses_fpu is a flag. Nonzero value indicates that the task will use the floating point unit.

signal is a function that is called, within the task environment and with interrupts disabled, when the task becomes the current running task after a context switch.

The newly created real time task is initially in a suspend state. It is can be made active either with rt_task_make_periodic, rt_task_make_periodic_relative_ns or rt_task_resume.

On multiprocessor systems rt_task_init_cpuid assigns task to a specific CPU *cpuid*. rt_task_init automatically selects which CPU the task will run on. This assignment may be changed by calling rt_set_runnable_on_cpus or rt_set_runnable_on_cpuid. If *cpuid* is invalid rt_task_init_cpuid falls back to automatic CPU selection

Return Value

- 0 – Success
- -EINVAL – Task structure pointed by *task* is already in use
- -ENOMEM – *stack_size* bytes could not be allocated for the stack

int rt_task_delete (RT_TASK **task*);

Deletes a real time task previously created by rt_task_init or rt_task_init_cpuid.

task is the pointer to the task structure.

If task *task* was waiting for a semaphore it is removed from the semaphore waiting queue. Otherwise, any other task blocked on message exchange with *task* is unblocked.

Return Value

- 0 – Success
- -EINVAL – *task* does not refer to a valid task

int rt_task_make_periodic (RT_TASK **task*, RTIME *start_time*, RTIME *period*);

int rt_task_make_periodic_relative_ns (RT_TASK **task*, RTIME *start_delay*, RTIME *period*);

Mark the task *task*, previously created with rt_task_init, as suitable for a periodic execution, with period *period*, when rt_task_wait_period is called.

The time of first execution is given by *start_time* or *start_delay*. *start_time* is an absolute value measured in clock ticks. *start_delay* is relative to the current time and measured in nanosecs.

Return Value

- 0 – Success
- -EINVAL – *task* does not refer to a valid task

void rt_task_wait_period (void);

Suspends the execution of the currently running real time task until the next period is reached. The task must have been previously marked for execution with rt_task_make_periodic or rt_task_make_periodic_relative_ns.

void rt_task_yield (void);

Blocks the current task and puts it at the end of the list of ready tasks with the same priority. The scheduler makes the next ready task of the same priority active.

int rt_task_suspend (RT_TASK **task*);

Suspends execution of the task *task*. It will not be executed until a call to rt_task_resume or rt_task_make_periodic is made.

Return Value

- 0 – Success
- -EINVAL – *task* does not refer to a valid task

int rt_task_resume (RT_TASK **task*);

Resumes execution of the task *task* previously suspended by rt_task_suspend or makes a newly created task ready to run.

Return Value

- 0 – Success
- -EINVAL – *task* does not refer to a valid task

int rt_get_task_state (RT_TASK **task*);

Returns the state of a real time task. *task* is a pointer to the task structure.

Return Value: Task state is formed by the bitwise OR of one or more of the following flags:

READY	Task *task* is ready to run (i.e., unblocked).
SUSPENDED	Task *task* is suspended.
DELAYED	Task *task* is waiting for its next running period or expiration of a timeout.
SEMAPHORE	Task *task* is blocked on a semaphore.
SEND	Task *task* sent a message and is waiting for the receiver task.
RECEIVE	Task *task* is waiting for an incoming message.
RPC	Task *task* sent a Remote Procedure Call and the receiver has not gotten it yet.
RETURN	Task *task* is waiting for reply to a Remote Procedure Call.

Note: the returned task state is only "approximate" information. Timer and other hardware interrupts may cause a change in the state of the queried task before the caller can evaluate the returned value. Caller should disable interrupts if it wants reliable info about another task.

RT_TASK *rt_whoami (void);

Calling rt_whoami allows a task to get a pointer to its own task structure.

Return Value: Pointer to the currently running task.

int rt_task_signal_handler (RT_TASK **task*, void (**handler*)(void));

Installs or changes the signal function of a real time task.

task is a pointer to the real time task

handler is the entry point of the signal function.

Signal handler is a function called within the task environment and with interrupts disabled, when the task becomes the current running task after a context switch. The signal handler function can also be set when the task is created with rt_task_init.

Return Value

- 0 – Success
- -EINVAL – *task* does not refer to a valid task

void rt_set_runnable_on_cpus (RT_TASK **task*, unsigned int *cpu_mask*);

void rt_set_runnable_on_cpuid (RT_TASK **task*, unsigned int *cpuid*);

Select one or more CPUs that are allowed to run task *task*. rt_set_runnable_on_cpuid assigns task to a specific CPU while rt_set_runnable_on_cpus magically selects one CPU from the given set that task *task* will run on. Bit<n> of cpu_mask enables CPU<n>.

If no CPU selected by *cpu_mask* or *cpuid* is available, both functions automatically select a possible CPU.

Note: This call has no effect on uniprocessor systems.

int rt_task_use_fpu (RT_TASK* *task*, int *use_fpu_flag*);

void rt_linux_use_fpu (int *use_fpu_flag*);

rt_task_use_fpu informs the scheduler that the real time task *task* will use floating point arithmetic operations.

rt_linux_use_fpu informs the scheduler that user space Linux processes will use floating point arithmetic operations.

If *use_fpu_flag* has nonzero value, FPU context is also switched when *task* or the kernel becomes active. This makes task switching slower. Initial value of this flag is set by rt_task_init when the real time task is created. By default Linux "task" has this flag cleared. It can be set with LinuxFpu command line parameter of the rtai_sched module.

Return Value (rt_task_use_fpu)

- 0 – Success

- -EINVAL – task does not refer to a valid task

void rt_preempt_always (int *yes_no*);

void rt_preempt_always_cpuid (int *yes_no*, unsigned *intcpu_id*);

In the one-shot mode a timed task is made active/current at the expiration of the timer shot. The next timer expiration is programmed by choosing among the timed tasks the one with a priority higher than the current after the current has released the CPU, always assuring the Linux timing. While this policy minimizes the programming of the one-shot mode, enhancing efficiency, it can be unsuitable when a task has to be guarded against looping by watch dog task with high priority value, as in such a case the latter as no chance of running.

Calling these functions with nonzero value assures that a timed high priority preempting task is always programmed to be fired while another task is current. The default is no immediate preemption in one-shot mode, firing of the next shot programmed only after the current task releases the CPU.

Initial value of this flag can be set with PreemptAlways command line parameter of the rtai_sched module.

Note: currently both functions are identical. Parameter *cpu_id* is ignored.

Timer Functions

void rt_set_oneshot_mode (void);

void rt_set_periodic_mode (void);

rt_set_oneshot_mode sets the timer to one-shot timing mode. It consists in a variable timing based on the cpu clock frequency. This allows task to be timed arbitrarily. It must be called before using any time related function, including conversions.

rt_set_periodic_mode sets the timer to periodic timing mode. It consists of a fixed frequency timing of the tasks in multiple of the period set with a call to start_rt_timer. The resolution is that of the 8254 frequency (1193180 hz). Any timing request not an integer multiple of the period is satisfied at the closest period tick. This is the default mode.

Note: Stopping the timer by stop_rt_timer sets the timer back into its default (periodic) mode. Call rt_set_oneshot_mode before each start_rt_timer if required.

RTIME start_rt_timer (int *period*);

void stop_rt_timer (void);

start_rt_timer starts the timer with a period *period*. The period is in internal count units and is required only for the periodic mode. In the one-shot the parameter value is ignored.

stop_rt_timer stops the timer. The timer mode is set to periodic

RTIME count2nano (RTIME *timercounts*);

RTIME nano2count (RTIME *nanosecs*);

count2nano converts the time of *timercounts* internal count units into nanoseconds.

nano2count converts the time of *nanosecs* nanoseconds into internal counts units.

Remember that the count units are related to the cpu frequency in one-shot mode and to the 8254 frequency (1193180 Hz) in periodic mode.

RTIME rt_get_time (void);

RTIME rt_get_time_ns (void);

RTIME rt_get_cpu_time_ns (void);

rt_get_time returns the number of real time clock ticks since RT_TIMER bootup (*whatever this means*). This number is multiple of the 8254 period in periodic mode, while is multiple of cpu clock period in one-shot mode.

rt_get_time_ns is the same as rt_get_time but the returned time is converted to nanoseconds.

rt_get_cpu_time_ns ???

RTIME next_period (void);

Returns the time when the caller task will run next. This is only relevant for periodic tasks.

void rt_busy_sleep (int *nanosecs*);

void rt_sleep (RTIME *delay*);

void rt_sleep_until (RTIME *time*);

rt_busy_sleep delays the execution of the caller task without giving back the control to the scheduler. This function burns up CPU cycles in a busy wait loop. It should be used for very short synchronization delays only.

nanosecs is the number of nanoseconds to wait.

rt_sleep suspends execution of the caller task for a time of *delay* internal count units. During this time the CPU is used by other tasks.

rt_sleep_until is similar to **rt_sleep** but the parameter time is the absolute time when the task should wake up. If the given *time* is already passed this call has no effect

Note: a higher priority task or interrupt handler can run during wait so the actual time spent in these functions may be longer than that specified.

Semaphore Functions

All of the core communication and synchronization mechanisms have a similar API.

#include "rtai_sem.h"

void rt_sem_init (SEM* *sem*, int *value*);

Initializes a semaphore *sem*. A semaphore can be used for communication and synchronization among real-time tasks. *sem* must point to a statically allocated structure. *value* is the initial value of the semaphore (usually 1). The initial value must be nonnegative.

A positive value of the semaphore variable shows how many tasks can do a wait operation without blocking. (Say how many tasks can enter the critical region.) A negative semaphore value shows how many tasks are blocked on it.

int rt_sem_delete (SEM* *sem*);

Deletes a semaphore previously created with rt_sem_create. *sem* points to the structure used in the corresponding call to rt_sem_create.

Any tasks blocked on this semaphore are allowed to run when the semaphore is deleted

Return Value

- 0 – Success
- SEM_ERR – sem does not refer to a valid semaphore. (-EINVAL would be a more consistent return value)

int rt_sem_signal (SEM* *sem*);

This is the semaphore post (sometimes known as *give*, *signal*, or *V*) operation. It is typically called when the task leaves a critical region. The semaphore value is incremented and tested. If the value is not positive, the first task in semaphore's waiting queue is allowed to run. rt_sem_signal does not block the caller task.

Return Value

- 0 – Success
- SEM_ERR – *sem* does not refer to a valid semaphore

int rt_sem_wait (SEM* *sem*);

This is the semaphore pend (sometimes known as *take*, *wait*, or *P*) operation. It is typically called when a task enters a critical region. The semaphore value is decremented and tested. If it is still nonnegative rt_sem_wait returns immediately. Otherwise the caller task is blocked and queued up. Queuing may happen in either priority order or FIFO order as determined by the compile time option SEM_PRIORD. In this case rt_sem_wait returns if

- The caller task is at the head of the waiting queue and another task issues a rt_sem_signal;
- An error occurs (e.g., the semaphore is destroyed);

Return Value

- On success an indeterminate value that somehow depends on the value of the semaphore is returned. This should be considered a bug
- SEM_ERR – *sem* does not refer to a valid semaphore

int rt_sem_wait_if (SEM* *sem*);

A version of rt_sem_wait that never blocks the caller. The return value indicates whether or not the calling task "got" the semaphore.

Return Value

- 0 – Semaphore was not available
- > 0 – "Previous" value of the semaphore. The semaphore has been decremented
- SEM_ERR – *sem* does not refer to a valid semaphore.

int rt_sem_wait_until (SEM* *sem*, RTIME *time*);

int rt_sem_wait_timed (SEM* *sem*, RTIME *delay*);

These are timed versions of rt_sem_wait. If the current semaphore value is less than 0 and the specified time interval expires before another task posts to the semaphore, these calls return with an error. rt_sem_wait_timed waits for up to *delay* internal counts. rt_sem_wait_until waits until an absolute time.

Return Value

- On success an indeterminate value that somehow depends on the value of the semaphore is returned. This should be considered a bug.
- SEM_ERR – *sem* does not refer to a valid semaphore.
- SEM_TIMEOUT – The specified interval expired before the semaphore became available.

Mailbox Functions

Mailboxes are a flexible method of task-to-task communication. Tasks are allowed to send arbitrary size messages by using any mailbox buffer size. Clearly you should use a buffer sized at least as big as the largest message you envisage. However if you expect a message larger than the average message size very rarely you can use a smaller buffer without much loss of efficiency. In such a way you can set up your own mailbox usage protocol, e.g., using fix size messages with a buffer that is an integer multiple of such a size guarantees that each message is sent/received atomically to/from the mailbox. Multiple senders and receivers are allowed and each will get the service it requires in turn, according to its priority.

#include "rtai_mbx.h"

int rt_mbx_init (MBX* *mbx*, int size);

Initializes a mailbox of size *size*. *mbx* points to a statically allocated mailbox structure. RTAI will dynamically allocate the buffer space.

Return Value

- 0 – Success
- -EINVAL – Space could not be allocated for the mailbox buffer

int rt_mbx_delete (MBX* *mbx*);

Removes a mailbox previously created with rt_mbox_init. *mbx* points to the structure used in the corresponding call to rt_mbox_init.

Return Value

- 0 – Success
- -EINVAL – *mbx* does not point to a valid mailbox

int rt_mbx_send (MBX* ***mbx***, void* ***msg***, int ***msg_size***);

Sends a message *msg* of *msg_size* bytes to the mailbox *mbx*. The caller will be blocked until the whole message is enqueued or an error occurs.

Return Value

- 0 – Success
- -EINVAL – *mbx* does not point to a valid mailbox

int rt_mbx_send_wp (MBX* ***mbx***, void* ***msg***, int ***msg_size***);

int rt_mbx_send_if (MBX* ***mbx***, void* ***msg***, int ***msg_size***);

Nonblocking versions of rt_mbx_send.

rt_mbx_send_wp sends as much as possible of message *msg* to mailbox *mbx* then returns immediately. "wp" means send "what's possible."

rt_mbx_send_if sends message *msg* to the mailbox *mbx* only if the entire message can be enqueued without blocking. Otherwise it returns an error.

Return Value

- >= 0 – Success. A nonzero value is the number of bytes of the message not sent.
- -EINVAL – *mbx* does not point to a valid mailbox

int rt_mbx_send_until (MBX* ***mbx***, void* ***msg***, int ***msg_size***, RTIME ***time***);

int rt_mbx_send_timed (MBX* ***mbx***, void* ***msg***, int ***msg_size***, RTIME ***delay***);

Timed versions of rt_mbx_send. These functions return after the specified time interval expires whether or not the entire message has been sent.

rt_mbx_send_until. waits until an absolute time.

rt_mbx_send_timed waits for up to *delay* internal counts.

Return Value

- >= 0 – A nonzero value is the number of bytes of the message *not* sent
- -EINVAL – *mbx* does not point to a valid mailbox

int rt_mbx_receive (MBX* *mbx*, void* *msg*, int *msg_size*);

Receives a message from the mailbox *mbx*. *msg* points to a buffer of *msg_size* bytes provided by the caller. The caller will be blocked until all bytes of the message arrive or an error occurs.

Return Value

- >= 0 – Number of bytes received from mailbox
- -EINVAL – *mbx* does not point to a valid mailbox

int rt_mbx_receive_wp (MBX* *mbx*, void* *msg*, int *msg_size*);

int rt_mbx_receive_if (MBX* *mbx*, void* *msg*, int *msg_size*);

Nonblocking versions of rt_mbx_receive.

rt_mbx_receive_wp receives at most *msg_size* bytes of message from mailbox *mbx* then returns immediately.

rt_mbx_receive_if receives a message from the mailbox *mbx* if the whole message of *msg_size* bytes is available immediately.

Return Value

- >= 0 – Number of bytes received from mailbox
- -EINVAL – *mbx* does not point to a valid mailbox

int rt_mbx_receive_until (MBX* *mbx*, void* *msg*, int *msg_size*, RTIME *time*);

int rt_mbx_receive_timed (MBX* *mbx*, void* *msg*, int *msg_size*, RTIME *delay*);

Timed versions of rt_mbx_receive. These functions return after the specified time interval expires whether or not the entire message has been sent.

rt_mbx_receive_until. waits until an absolute time.

rt_mbx_receive_timed waits for up to *delay* internal counts.

Return Value

- >= 0 – Number of bytes received from mailbox
- -EINVAL – *mbx* does not point to a valid mailbox

Message Handling Functions

This set of functions implements a direct task-to-task messaging mechanism. The message content is restricted to a single integer.

#include “rtai_msg.h”

RT_TASK* rt_send (RT_TASK* *task*, unsigned int *msg*);

RT_TASK* rt_send_if (RT_TASK* *task*, unsigned int *msg*);

Sends the message *msg* to the task *task*. rt_send blocks the calling task until the destination task, *task*, gets the message. If multiple tasks are sending messages to *task*, they are queued in either priority order or FIFO order as determined by compile time option MSG_PRIORD. rt_send_if doesn’t block.

Return Value

- Success – *task* (the pointer to the task that received the message)
- 0 – rt_send: *task* was killed before it could receive the message
  rt_send_if: *task* was not ready to receive the message
- MSG_ERR – *task* does not reference a valid task.

RT_TASK* rt_send_until (RT_TASK* *task*, unsigned int *msg*, RTIME *time*);

RT_TASK* rt_send_timed (RT_TASK* *task*, unsigned int *msg*, RTIME *delay*);

Timed versions of rt_ send. These functions return after the specified time interval expires whether or not *msg* was successfully sent.

Return Value

- Success – *task* (the pointer to the task that received the message)
- 0 – Operation timed out. Message was not delivered
- MSG_ERR – *task* does not reference a valid task

RT_TASK* rt_receive (RT_TASK* *task*, unsigned int **msg*);

RT_TASK* rt_receive_if (RT_TASK* *task*, unsigned int **msg*);

Gets a message from the task specified by *task*. If *task* is equal to 0, the caller accepts a message from any task. If there is a pending message, rt_receive returns immediately. Otherwise the caller task is blocked and queued up. (Queueing may happen in priority order or on FIFO base. This is determined by compile time option MSG_PRIORD.) *msg* points to a buffer provided by the caller. rt_receive_if returns immediately whether or not a message is pending.

Return Value

- Success – *task* (the pointer to the task that sent the message)
- 0 – rt_receive: *task* was killed before it could send the message.
  rt_receive_if: no message was sent
- MSG_ERR – *task* does not reference a valid task

RT_TASK* rt_receive_until (RT_TASK* task, unsigned int **msg*, RTIME *time*);

RT_TASK* rt_receive_timed (RT_TASK* task, unsigned int **msg*, RTIME *delay*);

Timed versions of rt_receive. These functions return after the specified time interval expires whether or not *msg* was successfully sent.

Return Value

- Success – *task* (the pointer to the task that received the message)
- 0 – Operation timed out. Message was not received
- MSG_ERR – *task* does not reference a valid task

Remote Procedure Calls

Although this mechanism is called "Remote Procedure Calls," in reality it is just a full duplex version of the message handling described above.

#include "rtai_msg.h"

RT_TASK *rt_rpc (RT_TASK **task*, unsigned int *msg*, unsigned int **reply*);

RT_TASK *rt_rpc_if (RT_TASK **task*, unsigned int *msg*, unsigned int **reply*);

Makes a Remote Procedure Call. RPC is like a send/receive pair. rt_rpc sends the message *msg* to the task *task* then waits until a reply is received. The caller task is always blocked and queued up. (Queuing may happen in priority order or on FIFO base. This is determined by compile time option MSG_PRIORD.) The receiver task may get the message with any rt_receive_* function. It can send the answer with rt_return. *reply* points to a buffer provided by the caller. rt_rpc_if doesn't block.

Return Value

- Success – *task* (the pointer to the task that received the message)
- 0 – *task* was not ready to receive the message (rt_rpc_if) or receiver task was killed before receiving the message
- MSG_ERR – *task* does not reference a valid task

RT_TASK *rt_rpc_until (RT_TASK **task*, unsigned int *msg*, unsigned int **reply*, RTIME *time*);

RT_TASK *rt_rpc_timed (RT_TASK **task*, unsigned int *msg*, unsigned int **reply*, RTIME *delay*);

Timed versions of rt_rpc. These functions return after the specified time interval expires whether or not *msg* was successfully sent.

Return Value

- Success – *task* (the pointer to the task that received the message)
- 0 – Operation timed out. Message was not delivered
- MSG_ERR – *task* does not reference a valid task

RT_TASK *rt_return (RT_TASK **task*, unsigned int *result*);

Sends *result* back to *task*. If the task calling rt_rpc_* previously is not waiting for the answer (i.e., killed or timed out) this return message is silently discarded.

Return Value

- Success – *task* (the pointer to the task that received the reply)
- 0 – Reply was not delivered
- MSG_ERR – *task* does not reference a valid task

int rt_isrpc (RT_TASK **task*);

After receiving a message, a task can determine whether the sender *task* is waiting for a reply or not by calling rt_isrpc. No answer is required if the message sent by a rt_send_* function or the sender called rt_rpc_timed or rt_rpc_until but it is already timed out.

Return Value

- 0 – *msg* was sent by rt_send_*, no reply necessary
- Nonzero – *task* is expecting a reply

rt_isrpc is not necessary because rt_return is smart enough to determine if a reply is required. Use of rt_isrpc is discouraged.

Module rtai_fifos

rtai_fifos provides a point-to-point sequenced mechanism for communicating between Kernel Space real-time tasks and User Space processes. The Kernel Space API is described here. User Space processes treat RTAI FIFOs as character devices /dev/rtfn. These devices are accessed from User Space using the normal open(), read() and write() system calls.

Reading and writing to FIFOs in Kernel Space is nonblocking.

#include <rtai_fifos.h>

int rtf_create (unsigned int *fifo*, int *size*);

Creates a real-time FIFO (RT-FIFO) of initial size *size* and assigns it the identifier *fifo*.

fifo is a small positive integer that identifies the FIFO on further operations. It must be less than RTF_NO.

fifo may refer an existing RT-FIFO. In this case the size is adjusted if necessary.

Return Value

- Success – *size* (the argument passed to rtf_create)
- -ENODEV – *fifo* is greater than or equal to RTF_NO
- -ENOMEM – *size* bytes could not be allocated for the RT-FIFO

int rtf_destroy (unsigned int *fifo*);

Closes a real-time *fifo* previously created/reopened with rtf_create() or rtf_open_sized(). An internal mechanism counts how many times a *fifo* was opened. Opens and closes must be in pair. rtf_destroy() should be called as many times as rtf_create() was. After the last close the *fifo* is really destroyed.

Return Value

- Success – The number of real-time tasks still having this FIFO open. Zero means the FIFO really was closed.
- -ENODEV – *fifo* is greater than or equal to RTF_NO
- -EINVAL – *fifo* is not a valid open FIFO

int rtf_reset (unsigned int *fifo*);

Removes any data posted to, but not yet removed from, *fifo*. The successful result is that *fifo* contains no data.

Return Value

- Success – 0
- -ENODEV – *fifo* is greater than or equal to RTF_NO
- -EINVAL – *fifo* is not a valid open FIFO

int rtf_resize (unsigned int *fifo*, int *size*);

Modifies the real-time FIFO *fifo*, previously created with rtf_create(), to have a new size of *size*. Any data currently in *fifo* is discarded.

Return Value

- Success – *size* (the argument passed to rtf_resize)
- -ENODEV – *fifo* is greater than or equal to RTF_NO
- -EINVAL – *fifo* is not a valid open FIFO
- -ENOMEM – *size* bytes could not be allocated for the RT-FIFO

int rtf_put (unsigned int *fifo*, void **buf*, int *count*);

Write a block of data to a real-time FIFO previously created with rtf_create. *fifo* is the ID with which the RT-FIFO was created. *buf* points the block of data to be written. *count* is the size of the block in bytes. This mechanism is available only to real-time tasks; Linux processes use a write() to the corresponding /dev/fifo<*n*> device to enqueue data to a FIFO.

Return Value

- Success – the number of bytes written. Note that this value may be less than *count* if *count* bytes of free space is not available in the FIFO
- -ENODEV – *fifo* is greater than or equal to RTF_NO
- -EINVAL – *fifo* is not a valid open FIFO

int rtf_get (unsigned int *fifo*, void **buf*, int *count*);

Reads a block of data from a real-time FIFO previously created with a call to rtf_create. *fifo* is the ID with which the RT-FIFO was created. *buf* points a buffer of *count* bytes size provided by the caller. This mechanism is available only to real-time tasks; Linux processes use a **read()** from the corresponding FIFO device to dequeue data from a FIFO.

Return Value

- Success – the number of bytes read. Note that this value may be less than *count* if *count* bytes were not available in the *fifo* at the time of the call
- -ENODEV – *fifo* is greater than or equal to RTF_NO.
- -EINVAL – *fifo* is not a valid open FIFO

int rtf_create_handler (unsigned int *fifo*, int (**handler*)(unsigned int *fifo*));

int rtf_create_handler (unsigned int *fifo*, X_FIFO_HANDLER(*handler*));

Installs a handler that is executed when data is written to or read from a real-time FIFO. *fifo* is an RT-FIFO that must have previously been created with a call to **rtf_create**. The function pointed by *handler* is called whenever a Linux process accesses that FIFO. The X_FIFO_HANDLER form allows for an extended handler function prototyped as:

int (**handler*)(unsigned int *fifo*, int *rw*)

This allows the handler to determine whether it was called as the result of a read (rw = 'r') or a write (rw = 'w').

rtf_create_handler is often used in conjunction with **rtf_get** to process data acquired asynchronously from a Linux process. The installed handler calls **rtf_get** when data is present. Because the handler is only executed when there is activity on the FIFO, polling is not necessary.

Return Value

- Success – 0
- -EINVAL – *fifo* is not a valid open FIFO

Module rtai_shm

rtai_shm supports memory regions shared between Kernel Space RTAI tasks and User Space processes.

#include <rtai_shm.h>

int rtheap_init (rtheap_t **heap*, void **heapaddr*, u_long *heapsize*, u_long *pagesize*)

Initializes a memory heap suitable for dynamic memory allocation requests. *heap* is the address of a heap descriptor the memory manager will use to store the allocation data. *heapaddr* points to a statically defined heap storage area. *heapsize* is the length of the storage area. *pagesize* is the size in bytes of chunks into which the storage area is divided. In the current implementation it must be a power of two.

If *heapaddr* is NULL, the heap manager attempts to extend the heap whenever it runs out of memory. In this case, *heapsize* is the size of the extent that will be requested from Linux. If this dynamically extendable mode is used, then the allocation function, rtheap_alloc, is considered to be "unbounded" because we don't know how long Linux might take to allocate an extent. If the heap is statically allocated, the allocation function is deterministic.

Return Value

- 0 – success
- *RTHEAP_PARAM* – a parameter is invalid
- *RTHEAP_NOMEM* – no initial extent can be allocated for a dynamically extendable heap

void rtheap_destroy (rtheap_t **heap*)

Destroys a memory heap created by rtheap_init. Dynamically allocated extents are returned to Linux.

void *rtheap_alloc (rtheap_t **heap*, u_long *size*, int *mode*)

Allocates a contiguous memory region of *size* bytes from an active memory *heap*. *size* is rounded up to the *pagesize* specified to rtheap_init. *mode* is a set of flags affecting the operation.

Returns NULL if the memory can't be allocated.

int rtheap_free (rtheap_t **heap*, void **block*)

Releases a memory *block* back to the memory *heap* it was previously allocated from.

Return Value

- 0 – success
- *RTHEAP_PARAM* – *block* does not belong to the specified *heap*

void *rt_shm_alloc (unsigned long *name*, int *size*, int *suprt*);

Allocates a shared memory region named *name* with *size* size bytes that can be shared between Kernel and User space. The first call to rt_shm_alloc for a given *name* allocates the space. Subsequent calls with the same name, whether from Kernel Space or User Space, just attach to it and increase the usage count. *suprt* is the kernel allocation method to be used, it can be:

- *USE_VMALLOC* – use vmalloc;
- *USE_GFP_KERNEL* – use kmalloc with *GFP_KERNEL*;
- *USE_GFP_ATOMIC* – use kmalloc with *GFP_ATOMIC*;
- *USE_GFP_DMA* – use kmalloc with *GFP_DMA*.

Return Value

- Success – Address of allocated space
- 0 – Unable to allocate requested space

void rt_shm_free (unsigned long *name*);

Frees the shared memory region identified by *name*. Actually, the region is just unmapped until the last process/task attached to this region calls rt_shm_free. Then the memory space is freed.

Legacy Macros

These macros define the older form of shared memory using the functions above. They should not be used in new applications.

Kernel space

```
#define rtai_kmalloc(name, size) rt_shm_alloc(name, size, USE_VMALLOC)
#define rtai_kfree(name) rt_shm_free(name)
```

User space

```
#define rtai_malloc(name, size) _rt_shm_alloc(0, name, size, USE_VMALLOC, 0)
#define rtai_free(name, adr) rt_shm_free(name)
```

Utility Name Functions

These functions are also used by LXRT.

#include “rtai_nam2num.h”

unsigned long nam2num (const char **name*);

Translates an ASCII *name* of up to six characters to an unsigned long id. The valid character set is:

'A' to 'Z', 'a' to 'z' (case not preserved)
'0' to '9', '_'

All other characters are translated to a value that is converted back to '$' by num2nam.

Return Value

- Success – the converted id
- There is no error condition

void num2nam (unsigned long *id*, char **name*);

Translates *id* back to an ASCII *name* string using the same algorithm as nam-2num.

Module rtai_lxrt

#include <rtai_lxrt.h>

Allows the RTAI API to be used (almost) transparently from User Space processes. This section describes only the functions that differ from the Kernel Space API. All other functions are used as described in previous sections.

Object Initialization

RT_TASK *rt_task_init (unsigned int *id*, int *priority*, int *stack_size*, int *max_msg¬size*);

Creates a new real-time task in User Space named *id*, with priority *priority*. *stack_size*, and *max_msgsize* may be 0 in which case default values are used. Default stack size is 512 bytes, default max message size is 256 bytes.

Return Value

- Success – pointer to a task structure in Kernel Space. Note that this value must not be used directly in User Space. It may only be passed as an argument to other LXRT functions.
- 0 – Unable to create buddy task or *id* is already registered in the name space

#define rt_sem_init (id, *value*) rt_typed_sem_init(id, value, CNT_SEM)

Allocates and initializes a semaphore with name *id* and initial count *value*.

#define rt_mbx_init (*id*, *size*) rt_typed_mbx_init(id, size, FIFO_Q)

Allocates and initializes a mailbox with name *id* and size *size*.

#define rt_cond_init(*id*) rt_typed_sem_init(id, 0, BIN_SEM)

Allocates and initializes a conditional variable.

Kernel Space Namespace Utilities

Functions to manage the LXRT namespace from kernel modules.

#include "registry.h".

int rt_register (unsigned long *id*, void **adr*, int *typ*, struct task_struct **tsk*));

Associates the name *id* with the Kernel Space object pointed to by *adr* and registers this object in the LXRT namespace. This allows User Space LXRT processes to reference the object. The use of *typ* and *tsk* are not specified.

Return Value

- Success – positive integer. Actually the slot number in the namespace table
- 0 – Unable to register the object. Namespace table is full

int rt_drg_on_adr (void **adr*);

int rt_drg_on_name (unsigned long *id*);

Remove an object from the namespace. The object to be removed can be referenced either by address *(adr)* or name *(id)*.

Return Value

- Success – positive integer. Actually the slot number in the namespace table that the object occupied.
- 0 – Unable to de-register the object. Not found in namespace table

User Space Namespace Utilities

These functions allow User Space LXRT processes to access Kernel Space objects.

void *rt_get_adr (unsigned long *id*);

Returns the address of the object named *id*.

unsigned long rt_get_name (void **adr*);

Returns the name of the object at address *adr*.

Return Value

- Both functions return zero if the requested object is not in the namespace.

User Space Hard Real-Time

void rt_make_hard_real_time (void);

Gives a User Space process hard real-time characteristics. The process must be locked in memory. Hard real-time processes should avoid making any Linux system call that could cause a task switch. Note that only processes run by the root user can lock memory (however see rt_allow_nonroot_hrt() below).

void rt_make_soft_real_time (void);

Returns a process back to normal soft real-time behavior.

void rt_allow_nonroot_hrt (void);

Allows a nonroot user to lock memory and invoke hard real-time.

APPENDIX **B**

Posix Threads (Pthreads) Application Programming Interface

This is a summary of the features of Posix 1003.1c-1995 supported under RTAI.

Unless indicated otherwise, all Pthreads functions return an integer status code where zero indicates that the function succeeded and a negative number indicates an error condition. Pthreads makes a distinction between two categories of errors:

1. Mandatory ("if occurs") errors involve circumstances beyond the control of the programmer. You wouldn't be expected to know, for example, or even be able to determine, that there isn't sufficient virtual memory to create a new thread. So the system must always detect and report this kind of error.
2. Optional ("if detected") errors are conditions that are usually your mistake. Attempting to lock a mutex that hasn't been initialized or trying to unlock a mutex that is locked by another thread are examples of this category. It may simply be too expensive in terms of processor time or other system resources to detect and report all of these potential error conditions. A competent programmer should be able to detect and solve these problems without help from the system.

Some systems may provide a "debugging" mode that turns on some or all of these optional errors while you're developing code. When production code is ready for release, you turn off the debugging mode.

In the following descriptions, mandatory error conditions are indicated in **bold**. For function arguments that take symbolic values, the list of valid symbols is given and the default value is indicated in **bold**.

> **Note:** *The function names given here are the standard ones. Except for message queues, the RTAI implementation appends a "_rt" to all Pthreads function names.*

#include <pthread.h> // for User Space pthreads

#include <rtai_posix.h> // for Kernel Space pthreads using RTAI

Threads

int pthread_create (pthread_t **tid*, const pthread_attr_t **attr*, void *(*start) (void *), void **arg*);

Creates a new thread with optional creation attributes *attr*. The new thread executes the function *start* with argument *arg*. *tid* is the thread's ID or handle.

Errors

- EINVAL – *attr* is not a thread attribute object
- EAGAIN – insufficient resources available to create the thread

int pthread_exit (void **ret_val*);

Terminates the calling thread after first calling any registered cleanup handlers. *ret_val* is returned to any thread joining this one.

pthread_t pthread_self (void);

Returns the ID of the calling thread.

int pthread_equal (pthread_t *t1*, pthread_t *t2*);

Returns zero if *t1* and *t2* are the same thread. Returns nonzero otherwise. Useful for comparing pthread_self() against another stored thread identifier.

int pthread_yield (void);

Yields the processor and makes the thread ready only after all other ready threads at this priority level have run.

Errors

- ENOSYS – sched_yield not supported.

int pthread_join (pthread_t *thread*, void ***value*);

Wait for the specified thread to terminate and get its return value.

Errors

- EINVAL – thread is not joinable
- ESRCH – specified thread not found

- EDEADLK – attempt to join with self

int pthread_cancel (pthread_t *thread*);

Request that the specified thread be cancelled. Note that cancellation is asynchronous. Use pthread_join() to wait for a thread to terminate.

Errors

- ESRCH – specified thread not found

void pthread_testcancel (void);

Creates a deferred cancellation point in the calling thread. If another thread has requested to cancel this one, its termination process will be started.

Thread Attributes

int pthread_attr_init (pthread_attr_t **attr*);

Initializes a pthread attributes object with default values.

Errors

- **ENOMEM** – insufficient memory for attribute object

int pthread_attr_destroy (pthread_attr_t **attr*);

Destroys an attribute object. Note that this does not affect threads previously created using *attr*.

Errors

- EINVAL – *attr* is not a thread attribute object

int_pthread_attr_getstacksize (const pthread_attr_t **attr*, size_t **stacksize*);

int_pthread_attr_setstacksize (pthread_attr_t **attr*, size_t *stacksize*);

Gets or sets the stack size for threads created with this *attr*.

Errors

- **ENOSYS** – stack size not supported
- **EINVAL** – *stacksize* too small or too big (set only)
- EINVAL – *attr* is not a thread attribute object

Scheduling Policy Attributes

int pthread_attr_getschedparam (const pthread_attr_t **attr*, struct sched_param **sched_p*);

int pthread_attr_setschedparam (pthread_attr_t **attr*, struct sched_param **sched_p*);

Gets or sets the scheduling parameters used by threads created with this *attr*. Contents of *sched_p* are implementation-dependent.

Errors

- **ENOSYS** – priority scheduling is not supported
- EINVAL – *attr* is not a thread attribute object or *sched_p* invalid (set only)

int pthread_attr_getschedpolicy (const pthread_attr_t **attr*, int **policy*);

int pthread_attr_setschedpolicy (pthread_attr_t **attr*, int *policy*);

Gets or sets the scheduling policy used by threads created with this *attr*.

Policy = SCHED_FIFO

SCHED_RR

SCHED_OTHER

Default value is implementation-dependent

Errors

- **ENOSYS** – priority scheduling is not supported
- EINVAL – *attr* is not a thread attribute object or *policy* invalid (set only)

Mutexes

int pthread_mutex_init (pthread_mutex_t **mutex*, const pthread_mutexattr_t **mutex_attr*);

Creates a ***mutex*** object. Mutex attributes are not supported in RTAI.

Errors

- **EAGAIN** – insufficient resources other than memory
- **ENOMEM** – insufficient memory for mutex object

- **EPERM** – no privilege to perform this operation
- EBUSY – *mutex* is already initialized
- EINVAL – *mutex_attr* is not a mutex attribute object

int pthread_mutex_destroy (pthread_mutex_t **mutex*);

Destroys an existing mutex that is no longer needed.

Errors

- EBUSY – *mutex* is in use
- EINVAL – *mutex* is not a mutex

int pthread_mutex_lock (pthread_mutex_t **mutex*);

Locks a mutex. If *mutex* is already locked, the calling thread is blocked until *mutex* is subsequently unlocked. On return the calling thread "owns" *mutex* until it calls pthread_mutex_unlock().

Errors

- **EINVAL** – thread priority exceeds *mutex* priority ceiling
- EINVAL – *mutex* is not a mutex object
- EDEADLK – calling thread already owns *mutex*

int pthread_mutex_trylock (pthread_mutex_t **mutex*);

Locks *mutex* if it is currently unlocked. If *mutex* is locked return immediately with error code. This is a nonblocking method of locking a mutex.

Errors

- **EINVAL** – thread priority exceeds *mutex* priority ceiling
- **EBUSY** – *mutex* is already locked
- EINVAL – *mutex* is not a mutex object
- EDEADLK – calling thread already owns *mutex*

int pthread_mutex_unlock (pthread_mutex_t *mutex);

Unlocks *mutex*. If any threads are waiting on this mutex, one of them is awakened and becomes the new owner. The order in which threads are awakened depends on scheduling policy.

Errors

- EINVAL – *mutex* is not a mutex
- *EPERM* – calling thread does not own *mutex*

Mutex Attributes (not implemented in RTAI. Shown here for completeness)

int pthread_mutexattr_init (pthread_mutexattr_t **mutex_attr*);

Initializes a mutex attributes object with default values.

Errors

- **ENOMEM** – insufficient memory for attribute object

int pthread_mutexattr_destroy (pthread_mutexattr_t **mutex_attr*);

Destroys a mutex attribute object.

Errors

- EINVAL – mutex_attr is not a mutex attribute object

int pthread_mutexattr_getkind_np (const pthread_mutexattr_t **mutex_attr*, int **kind*);

int pthread_mutexattr_setkind_np (pthread_mutexattr_t **mutex_attr*, int *kind*);

Gets or sets the mutex "kind" or type. This is a nonportable Linux extension. See Chapter 12 for details on mutex kind.

kind = **PTHREAD_MUTEX_FAST_NP**
PTHREAD_MUTEX_RECURSIVE_NP
PTHREAD_MUTEX_ERRORCHECK_NP

Errors

- EINVAL – *mutex_attr* is not a mutex attribute object
- **EINVAL** – *kind* is invalid (set only)

int pthread_mutexattr_getprioceiling (const pthread_mutexattr_t **mutex_attr*, int **prioceiling*);

int pthread_mutexattr_setprioceiling (pthread_mutexattr_t **mutex_attr*, int *prioceiling*);

Gets or sets the priority ceiling at which threads run while owning a mutex created with *attr*.

Errors

- ENOSYS – Priority scheduling is not supported
- EINVAL – *attr* is not a mutex attribute object or *prioceiling* invalid (set only)
- *ENOPERM* – no permission to set *prioceiling*

int pthread_mutexattr_getprotocol (const pthread_mutexattr_t **mutex_attr*, int **protocol*);

int pthread_mutexattr_setprotocol (pthread_mutexattr_t **mutex_attr*, int *protocol*);

Gets or sets the mutex protocol for dealing with priority inversions.

protocol = ***PTHREAD_PRIO_NONE***
PTHREAD_PRIO_INHERIT
PTHREAD_PRIO_PROTECT

Errors

- ENOSYS – Priority scheduling is not supported
- EINVAL – *attr* is not a mutex attribute object or *protocol* invalid (set only)
- *ENOTSUP* – *protocol* value is not supported

Condition Variables

int pthread_cond_init (pthread_cond_t **cond*, const pthread_condattr_t **cond_attr*);

Creates a condition variable object, *cond*. Condition variable attributes are not supported under RTAI.

Errors

- EAGAIN – insufficient resources other than memory
- ENOMEM – insufficient memory for conditional variable object
- EBUSY – *cond* is already initialized

- EINVAL – *cond_attr* is not a condition variable attribute object

int pthread_cond_destroy (pthread_cond_t **cond*);

Destroys an existing *cond* that is no longer needed.

Errors

- EBUSY – *cond* is in use
- EINVAL – *cond* is not a condition variable

int pthread_cond_wait (pthread_cond_t **cond*, pthread_mutex_t **mutex*);

Waits on a condition variable *cond* until awakened either by signal or broadcast. ***mutex*** is unlocked (before wait) and relocked (after wait) inside pthread_cond_wait().

Errors

- EINVAL – *cond* or ***mutex*** is not valid
- EINVAL – different mutexes for concurrent waits
- EINVAL – ***mutex*** is not owned by calling thread

int pthread_cond_timedwait (pthread_cond_t **cond*, pthread_mutex_t **mutex*, const struct timespec **time*);

Waits on a condition variable *cond* until awakened either by signal or broadcast or until the absolute time *time* is reached. *mutex* is unlocked (before wait) and relocked (after wait) inside pthread_cond_wait().

Errors

- **ETIMEOUT** – *time* has expired
- EINVAL – *cond*, ***mutex*** or *time* is not valid
- EINVAL – different mutexes for concurrent waits
- EINVAL – ***mutex*** is not owned by calling thread

int pthread_cond_signal (pthread_cond_t **cond*);

Signals condition variable *cond* waking up one waiting thread. The order in which threads wake up depends on the scheduling policy.

Errors

- EINVAL – *cond* is not a condition variable

int pthread_cond_broadcast (pthread_cond_t **cond*);

Like signal except that **all** waiting threads are awakened.

Errors

- EINVAL – *cond* is not a condition variable

Condition Variable Attributes (not implemented in RTAI. Shown here for completeness)

int pthread_condattr_init (pthread_condattr_t **cond_attr*);

Initializes a condition variable attributes object with default values.

Errors

- **ENOMEM** – insufficient memory for attribute object

int pthread_condattr_destroy (pthread_condattr_t **cond_attr*);

Destroys a condition variable attribute object.

Errors

- EINVAL – *cond_attr* is not a condition variable attribute object

Message Queues

#include <rtai_mq.h>

The descriptions in this section do not distinguish between required and optional error codes. They list all of the error codes returned by the RTAI implementation.

mqd_t mq_open (char **mq_name*, int *oflags*, mode_t *permissions*, struct mq_attr **mq_attr*);

Creates a new message queue name *mq_name* or opens an existing one for use by the calling thread. *oflags* controls the way the message is accessed and, if necessary, opened as follows:

O_RDONLY, O_WRONLY or O_RDWR – normal access control

O_NONBLOCK – don't block if the queue is full/empty

O_CREAT – create the queue if it doesn't already exist.

O_EXCL – when used with O_CREAT, return an error if the queue already exists *permissions* specifies the User/GroupOther, read/write/execute permissions for the queue. *mq_attr* specifies the "geometry" of the queue including:

mq_maxmsgs – maximum number of messages the queue can hold

mq_msgsize – maximum size of an individual message

mq_flags – blocking, nonblocking behavior (only used by mq_setattr() and mq_getattr())

mq_curmsgs – number of messages currently in the queue

permissions and *mq_attr* are only relevant if this call creates the message queue.

On success mq_open() returns a queue descriptor for use in subsequent message queue calls.

Errors

- ENOMEM – insufficient memory to create queue
- EMFILE – no message queue descriptors available
- EACCES – message queue exists and permissions in *oflags* denied or permission to create queue is denied.
- EEXIST – message queue already exists and O_CREAT and O_EXCL were specified
- EINVAL – *mq_name* or *mq_attr* invalid
- ENOENT – message queue doesn't exist and O_CREAT not specified
- ENAMETOOLONG – *mq_name* is too long

int mq_close (mqd_t *mq*);

Closes the calling thread's link to message queue *mq*. The message queue still exists and any messages currently posted there may be accessed by other threads linked to *mq*.

Errors

- EBADF – *mq* is not a valid message queue identifier

int mq_unlink (char **mq_name*);

Destroys the message queue *mq_name* but only if no other threads have an open link to it. The queue's memory is deallocated and any messages remaining are lost. If other threads have open links to this queue it is marked for later deletion when the last open link is closed. Once mq_unlink() is called on a queue no further links can be opened to it.

mq_unlink() returns 0 if it successfully destroys the queue. A positive return value is the number of links currently open to this queue.

Errors

- ENOENT – *mq_name* is not a valid message queue

int mq_send (mqd_t mq, const char **msg*, size_t *msglen*, unsigned int *prio*);

Sends *msg* of length *msglen* and priority *prio* to message queue *mq*. Messages are placed on the queue in priority order. Within the same priority they are posted in FIFO order.

Errors

- EBADF – *mq* is not a valid message queue identifier
- EINVAL – *prio* is greater than MQ_MAX_PRIO or the calling thread does not have proper queue access permissions.
- EMSGSIZE – *msglen* is greater than *mq_msgsize* for this queue
- EAGAIN – the queue is full and it is nonblocking

size_t mq_receive (mqd_t mq, char **msg_buff*, size_t *buflen*, unsigned int **prio*);

Receives a message from queue *mq* and puts it in *msg_buff* of length *buflen*. The message priority is returned in *prio*. A positive return value is the length of the received message.

Errors

- EBADF – *mq* is not a valid message queue identifier
- EINVAL – calling thread does not have proper queue access permissions
- EMSGSIZE – *buflen* is less than *mq_msgsize* for this queue
- EAGAIN – the queue is empty and it is nonblocking

int mq_notify (mqd_t mq, const struct sigevent **notify*);

#include <asm/siginfo.h> // struct sigevent

Allows the calling thread to arrange for asynchronous notification of the arrival of a message in *mq*. A message queue can only register one such notification request. A previously registered notification request can be removed by passing *notify* as NULL.

Errors

- EBADF – *mq* is not a valid message queue identifier
- –1 – notification request already registered with this queue or the request can't be cleared because it is owned by another thread.

int mq_getattr (mqd_t mq, struct mq_attr **mq_attr*);

Returns the attribute structure of message queue *mq*.

Errors

- EBADF – *mq* is not a valid message queue identifier

int mq_setattr (mqd_t mq, const struct mq_attr **new_attr*, struct mq_attr **old_attr*);

Sets the attributes of *mq* from *new_attr*. Only the *mq_flags* can be changed by this call. All others are unaffected. If *old_attr* is non-NULL, the original attribute structure of *mq* is stored here.

Errors

- EBADF – *mq* is not a valid message queue identifier
- EINVAL – calling thread does not have proper queue access permissions

APPENDIX C

Why Software Should Not Have Owners

—Richard Stallman, Free Software Foundation

Digital information technology contributes to the world by making it easier to copy and modify information. Computers promise to make this easier for all of us.

Not everyone wants it to be easier. The system of copyright gives software programs "owners," most of whom aim to withhold software's potential benefit from the rest of the public. They would like to be the only ones who can copy and modify the software that we use.

The copyright system grew up with printing—a technology for mass production copying. Copyright fit in well with this technology because it restricted only the mass producers of copies. It did not take freedom away from readers of books. An ordinary reader, who did not own a printing press, could copy books only with pen and ink, and few readers were sued for that.

Digital technology is more flexible than the printing press: when information has digital form, you can easily copy it to share it with others. This very flexibility makes a bad fit with a system like copyright. That's the reason for the increasingly nasty and draconian measures now used to enforce software copyright. Consider these four practices of the Software Publishers Association (SPA):

- Massive propaganda saying it is wrong to disobey the owners to help your friend.
- Solicitation for stool pigeons to inform on their coworkers and colleagues.
- Raids (with police help) on offices and schools, in which people are told they must prove they are innocent of illegal copying.

- Prosecution (by the US government, at the SPA's request) of people such as MIT's David LaMacchia, not for copying software (he is not accused of copying any), but merely for leaving copying facilities unguarded and failing to censor their use.

All four practices resemble those used in the former Soviet Union, where every copying machine had a guard to prevent forbidden copying, and where individuals had to copy information secretly and pass it from hand to hand as "samizdat". There is of course a difference: the motive for information control in the Soviet Union was political; in the US the motive is profit. But it is the actions that affect us, not the motive. Any attempt to block the sharing of information, no matter why, leads to the same methods and the same harshness.

Owners make several kinds of arguments for giving them the power to control how we use information:

- *Name calling.* Owners use smear words such as "piracy" and "theft," as well as expert terminology such as "intellectual property" and "damage," to suggest a certain line of thinking to the public—a simplistic analogy between programs and physical objects.

  Our ideas and intuitions about property for material objects are about whether it is right to *take an object away* from someone else. They don't directly apply to *making a copy* of something. But the owners ask us to apply them anyway.

- *Exaggeration.* Owners say that they suffer "harm" or "economic loss" when users copy programs themselves. But the copying has no direct effect on the owner, and it harms no one. The owner can lose only if the person who made the copy would otherwise have paid for one from the owner.

  A little thought shows that most such people would not have bought copies. Yet the owners compute their "losses" as if each and every one would have bought a copy. That is exaggeration—to put it kindly.

- *The law.* Owners often describe the current state of the law, and the harsh penalties they can threaten us with. Implicit in this approach is the suggestion that today's law reflects an unquestionable view of morality—yet at the same time, we are urged to regard these penalties as facts of nature that can't be blamed on anyone.

  This line of persuasion isn't designed to stand up to critical thinking; it's intended to reinforce a habitual mental pathway.

It's elementary that laws don't decide right and wrong. Every American should know that, forty years ago, it was against the law in many states for a black person to sit in the front of a bus; but only racists would say sitting there was wrong.

- *Natural rights*. Authors often claim a special connection with programs they have written, and go on to assert that, as a result, their desires and interests concerning the program simply outweigh those of anyone else—or even those of the whole rest of the world. (Typically companies, not authors, hold the copyrights on software, but we are expected to ignore this discrepancy.)

  To those who propose this as an ethical axiom—the author is more important than you—I can only say that I, a notable software author myself, call it bunk.

  But people in general are only likely to feel any sympathy with the natural rights claims for two reasons.

  One reason is an overstretched analogy with material objects. When I cook spaghetti, I do object if someone else eats it, because then I cannot eat it. His action hurts me exactly as much as it benefits him; only one of us can eat the spaghetti, so the question is, which? The smallest distinction between us is enough to tip the ethical balance.

  But whether you run or change a program I wrote affects you directly and me only indirectly. Whether you give a copy to your friend affects you and your friend much more than it affects me. I shouldn't have the power to tell you not to do these things. No one should.

  The second reason is that people have been told that natural rights for authors is the accepted and unquestioned tradition of our society.

  As a matter of history, the opposite is true. The idea of natural rights of authors was proposed and decisively rejected when the US Constitution was drawn up. That's why the Constitution only *permits* a system of copyright and does not *require* one; that's why it says that copyright must be temporary. It also states that the purpose of copyright is to promote progress—not to reward authors. Copyright does reward authors somewhat, and publishers more, but that is intended as a means of modifying their behavior.

  The real established tradition of our society is that copyright cuts into the natural rights of the public—and that this can only be justified for the public's sake.

- *Economics*. The final argument made for having owners of software is that this leads to production of more software.

  Unlike the others, this argument at least takes a legitimate approach to the subject. It is based on a valid goal—satisfying the users of software. And it is empirically clear that people will produce more of something if they are well paid for doing so.

  But the economic argument has a flaw: it is based on the assumption that the difference is only a matter of how much money we have to pay. It assumes that "production of software" is what we want, whether the software has owners or not.

  People readily accept this assumption because it accords with our experiences with material objects. Consider a sandwich, for instance. You might well be able to get an equivalent sandwich either free or for a price. If so, the amount you pay is the only difference. Whether or not you have to buy it, the sandwich has the same taste, the same nutritional value, and in either case you can only eat it once. Whether you get the sandwich from an owner or not cannot directly affect anything but the amount of money you have afterwards.

  This is true for any kind of material object—whether or not it has an owner does not directly affect what it *is*, or what you can do with it if you acquire it.

  But if a program has an owner, this very much affects what it is, and what you can do with a copy if you buy one. The difference is not just a matter of money. The system of owners of software encourages software owners to produce something—but not what society really needs. And it causes intangible ethical pollution that affects us all.

What does society need? It needs information that is truly available to its citizens—for example, programs that people can read, fix, adapt, and improve, not just operate. But what software owners typically deliver is a black box that we can't study or change.

Society also needs freedom. When a program has an owner, the users lose freedom to control part of their own lives.

And above all society needs to encourage the spirit of voluntary cooperation in its citizens. When software owners tell us that helping our neighbors in a natural way is "piracy," they pollute our society's civic spirit.

This is why we say that free software is a matter of freedom, not price.

The economic argument for owners is erroneous, but the economic issue is real. Some people write useful software for the pleasure of writing it or for admiration and love; but if we want more software than those people write, we need to raise funds.

For ten years now, free software developers have tried various methods of finding funds, with some success. There's no need to make anyone rich; the median US family income, around $35k, proves to be enough incentive for many jobs that are less satisfying than programming.

For years, until a fellowship made it unnecessary, I made a living from custom enhancements of the free software I had written. Each enhancement was added to the standard released version and thus eventually became available to the general public. Clients paid me so that I would work on the enhancements they wanted, rather than on the features I would otherwise have considered highest priority.

The Free Software Foundation (FSF), a tax-exempt charity for free software development, raises funds by selling GNU CD-ROMs, T-shirts, manuals, and deluxe distributions, (all of which users are free to copy and change), as well as from donations. It now has a staff of five programmers, plus three employees who handle mail orders.

Some free software developers make money by selling support services. Cygnus Support, with around 50 employees (when this article was written), estimates that about 15% of its staff activity is free software development—a respectable percentage for a software company.

Companies including Intel, Motorola, Texas Instruments and Analog Devices have combined to fund the continued development of the free GNU compiler for the language C. Meanwhile, the GNU compiler for the Ada language is being funded by the US Air Force, which believes this is the most cost-effective way to get a high quality compiler. (Air Force funding ended some time ago; the GNU Ada Compiler is now in service, and its maintenance is funded commercially.)

All these examples are small; the free software movement is still small, and still young. But the example of listener-supported radio in this country (the US) shows it's possible to support a large activity without forcing each user to pay.

As a computer user today, you may find yourself using a proprietary program. If your friend asks to make a copy, it would be wrong to refuse. Cooperation is more important than copyright. But underground, closet cooperation does not make for a good society. A person should aspire to live an upright life openly with pride, and this means saying “No” to proprietary software.

You deserve to be able to cooperate openly and freely with other people who use software. You deserve to be able to learn how the software works, and to teach your students with it. You deserve to be able to hire your favorite programmer to fix it when it breaks.

You deserve free software.

Updated: $Date: 2001/09/15 20:14:02 $ $Author: fsl $

APPENDIX D

Upgrading From Kernel 2.4 to 2.6

The substantial changes between the 2.4 and 2.6 series kernels mean that many Linux distributions based on 2.4 kernels need some updating in order to successfully build and execute the 2.6 kernel. So the first thing to do is check the revision level of various tools and utilities and, if necessary, download updated versions.

Required

These are the minimum version levels for tools required to build or run a 2.6 series kernel.

Package	Minimum Version	Where to get Latest version	How to find Version
Gnu C Compiler	**2.95.3**	http://gcc.gnu.org/	gcc --version
Gnu make	**3.78**	ftp://ftp.gnu.org/gnu/make/	make --version
binutils	**2.12**	ftp://ftp.kernel.org/pub/linux/devel/binutils/	ld -v
util-linux	**2.10o**	ftp://ftp.kernel.org/pub/linux/utils/util-linux/	fdformat --version
module-init-tools	**0.9.9**	www.kernel.org/pub/linux/kernel/people/rusty/modules/	depmod -V
procps	**2.0.9**	www.tech9.net/rml/procps/	ps --version

Filesystems

Most likely you're running either the efs2 or efs3 filesystem, both of which require that e2fsprogs be up to date. The other packages in this list are for filesystems that are not widely used, but are included for completeness.

Package	Minimum Version	Where to get Latest version	How to find Version
e2fsprogs	**1.29**	http://e2fsprogs.source-forge.net/	tune2fs
jfsutils	**1.0.14**	www-124.ibm.com/jfs/	fsck.jfs -V
reiserfsprogs	**3.6.3**	www.namesys.com/	reiserfsck -V 2>&1 | grep reiserfsprogs
xfsprogs	**2.1.0**	http://oss.sgi.com/ projects/xfs/	xfs_db -V
nfs-utils	**1.0.5**	http://nfs.sourceforge.net/	showmount --version
procps	**2.0.9**	www.tech9.net/rml/procps/	ps --version

Miscellaneous Utilities

These are utilities that you may need to update if you use them.

Package	Minimum Version	Where to get Latest version	How to find Version
pcmcia-cs	**3.1.21**	http://pcmcia-cs.sourceforge.net/	cardmgr -V
quota-tools	**3.09**	http://sourceforge.net/ projects/linuxquota/	quota -V
ppp	**2.4.0**	ftp://ftp.samba.org/pub/ ppp/	ppd --version
isdn4k-utils	**3.1pre1**	www.isdn4linux.de/swpat. html/	isdnctrl 2>&1 | grep version
oprofile	**0.5.3**	http://oprofile.sourceforge. net/	oprofiled –version

Additional Resources

http://kerneltrap.org/node/799 – How to Upgrade To The 2.6 Kernel

http://thomer.com/linux/migrate-to-2.6.html – Migrating to Linux Kernel 2.6

`post-halloween-2.6.txt` – This file is in the `/tools` directory on the CD-ROM accompanying this book. It explains some of the new functionality to be found in the 2.6 kernel and some "pitfalls" you may encounter.

Index

M

N

O

P

R

S

T

U

W